D1261206

THE LIBRARY OF POLITICAL ECONOMY

POLITICAL ECONOMY is the old name for economics. In the hands of the great classical economists, particularly Smith, Ricardo and Marx, economics was the study of the working and development of the economic system in which men and women lived. Its practitioners were driven by a desire to describe, to explain and to evaluate what they saw around them. No sharp distinction was drawn between economic analysis and economic policy nor between economic behaviour and its interaction with the technical, social and political framework.

The Library of Political Economy has been established to provide widely based explanations of economic behaviour in contemporary society.

In examining the way in which new patterns of social organization and behaviour influence the economic system and policies for combating problems associated with growth, inflation, poverty and the distribution of wealth, contributors stress the link between politics and economics and the importance of institutions in policy formulation.

This 'open-ended' approach to economics implies that there are few laws that can be held to with certainty and, by the same token, there is no generally established body of theory to be applied in all circumstances. Instead economics as presented in this library provides a way of ordering events which has constantly to be updated and modified as new situations develop. This, we believe, is its interest and its challenge.

Volumes in the Library

Dangerous Currents: The State of Economics – Lester Thurow
The Political Economy of Nationalism – Dudley Seers
Women's Claims: A Study in Political Economy – Lisa Peattie and Martin Rein
Urban Inequalities under State Socialism – Ivan Szelenyi
Social Innovation and the Division of Labour – Jonathan Gershuny

Social Innovation and the Division of Labour

Jonathan Gershuny

OXFORD UNIVERSITY PRESS
1983

Oxford University Press, Walton Street, Oxford OX2 6DP

London Glasgow New York Toronto
Delhi Bombay Calcutta Madras Karachi
Kuala Lumpur Singapore Hong Kong Tokyo
Nairobi Dar es Salaam Cape Town
Melbourne Auckland

and associated companies in
Beirut Berlin Ibadan Mexico City Nicosia

Oxford is a trade mark of Oxford University Press

Published in the United States
by Oxford University Press, New York

British Library Cataloguing in Publication Data
Gershuny, Jonathan
Social innovation and the division of labour. –
(The Library of political economy)
1. Social conditions   2. Politics and government
I. Title   II. Series
361.6'13'091722   HC53.2
ISBN 0-19-874131-6
ISBN 0-19-874130-8 Pbk

Set by Hope Services, Abingdon
and Printed in Great Britain
at the University Press, Oxford
by Eric Buckley
Printer to the University

Foreword

WHAT follows is one part of the first round of publications from a project on the impact of technical change on life-styles and economic structures (with particular reference to the future of work), which has been in progress at the Science Policy Research Unit for some three years. It has been produced alongside two other books, *The New Service Economy*,[1] written by the present author together with Ian Miles, and *Changing Times*[2] (which examines evidence on the change in patterns of time use and activities in the UK over the last fifty years), again jointly authored, by Graham Thomas and the present writer. This book is intended to outline, in the briefest possible way, the overall framework which has been developed for the project as a whole; much of the credit for what follows should go to the above-mentioned writers, as well as to our fourth collaborator, Chris Zmroczek.

The project has been funded by a number of different agencies. The Anglo-German Foundation and the German Marshall Fund of the United States each provided roughly half the costs up to September 1981, at which point The Joseph Rowntree Memorial Trust took over, providing a three year programme grant. The European Commission Project 'FAST' funded part of the preparation of the Service Economy book.

The next phase of the project involves three different sorts of activities. There will be a historical element, developing some detailed case-studies of the effects of some particular technical innovations (domestic laundry equipment, cinema and television), both on activity patterns and paid employment. There will be a futurological element, using our model of the impact of domestic technological innovation to examine the likely effects of new telecommunications and computing equipment, again both on employment and on life-styles. And we shall develop our work on time use, both continuing the analysis of the data we have already collected, and, it is hoped, carrying out a new national 'time budget survey' in the UK during 1983.

[1] Frances Pinter, London, 1983. [2] To be published by OUP.

Contents

CHAPTER 1

Introduction: On the Subject of this Book

1.1 Two Views of Technology

THIS book is about the future of work. It is concerned with three of the different ways in which work is distributed; with the division of labour between the various industries and occupations; with the division of labour between paid and unpaid work activities; and, to a more limited extent, with the division of labour between different sorts of people, particularly between men and women. Though its concern is with the future, the main substance of the book is historical; it deals with the development patterns of various European economies over the last two or three decades. This historical material is used in the formulation of a new theory about the relationship between technical change and the three aspects of the division of labour.

There are two different ways of looking at the relationship between technical change and economic and social structure. The first starts in the workplace. Technical change is seen as something embodied in new production technologies, new machines, new ways of organizing work. It lowers the costs of production of particular commodities, changes relative prices, and hence alters consumption patterns. Or an invention leads to new products, which develop new markets. In either case, this first view sees a clear pattern of cause and effect which moves *from* production *to* consumption; technology affects life-styles by changing the relative costs and performances of the set of items which is offered for sale to consumers.

The second, complementary, way of looking at technical change starts not in the workplace, but in the household. Let us assume that households have a certain range of needs, a set of 'service functions' that they wish to satisfy – food, shelter, domestic services, entertainment, transport, medicine, education, and, more distantly, government services, 'law and order' and defence. It is a commonplace of economics that as societies get richer they wish to change their distribution of resources among these functions, they devote a smaller proportion of their national incomes to satisfying the more basic needs, and a larger to the more sophisticated, luxury categories. Indeed this observation is the bedrock of the conventional explanation of the industrial division of labour. Less well integrated into economic thought, however, is that we may see, over time, change in the technical and organizational means of provision for some of these functions. The particular means that a household employs to satisfy its needs for a particular function – the 'mode of provision' for the function – determines the household's pattern of expenditure on final commodities. Over time, the relative desirabilities of two alternative modes of provision for a particular function may change – as when the cost of a particular item of final services rises relative to the price of a durable good which may be used to satisfy the same function – with the consequence that the household changes from one mode of provision to the other. This change in the mode of provision for particular functions (or, at the societal level, change in the distribution of modes of provision) will be referred to as 'social innovation' (though of course this term in normal usage covers a broader range of phenomena); it is at the heart of the second view of technical change, and is the subject of this book.

1.2 Familiar Revolutions

Things closest to us are least easily seen. The foregoing is a very abstract description of a very familiar phenomenon: the lifetimes of most of the readers of this book have seen revolutionary changes in the distribution of modes of provision for a wide range of final service functions.

Entertainment, for example, was once largely purchased in the form of a completed final service. It is now largely acquired through the medium of what is in effect household capital equipment – televisions, music centres, video recorders, and leisure goods such as boats and sports equipment. Transport, again, was at one time mostly purchased in the form of final services; now we buy cars (capital equipment), and petrol, and roads (through our taxes) – and produce the final services ourselves. And domestic services were once either purchased as finished commodities (i.e. the middle and upper classes had servants) or else acquired by long hours of non-mechanized unpaid work. Now the purchase of final domestic services is no longer a major element in our societies' final demand patterns, it has been replaced by mechanized, capital intensive, unpaid, domestic work (the consequences for domestic work time are a matter of dispute – we explore this issue in Chapter 9). These social innovations have had major impacts throughout the developed world.

Perhaps the most important of these effects have been the creation of new markets. Motor vehicles, consumer durables, and electronics – 'white goods' – were, arguably, the industries on which the economic expansion of the 1950s and 1960s was based. In general, the more successful these industries were, the more successful the national economy. And this growth of new markets had spin-off elsewhere in the economy. It led to intermediate and investment demand in the heavy engineering and metal manufacturing industries. It led to an expansion of the construction industry, which built the infrastructure (roads etc.) necessary to use the domestic capital goods, and provided the improvements in the housing stock needed to contain the new productive domestic activities. And the improvement in general welfare that emerged from the social innovations provided economic space for the transfer of resources to an increased collective provision of services, education, and increasing, medicine. In short, social innovation may be seen as a sort of motor of economic development.

This book will explore the theme of social innovation in a number of ways. It will develop a very simple model showing how households' changing modes of provision for particular functions impinge both on the industrial distribution of employment and on the household division of time between paid and unpaid work. It will demonstrate how conventional national income statistics can be used to give empirical estimates of the effect of social innovation on economic structure. But the ultimate intention is futurological; the theoretical and empirical arguments will be used to explore two important themes in the discussion of the prospects for work organization – 'informalization' and 'informatization'.

1.3 A Simple Model

This short book is an essay of persuasion, intended to convince readers of the importance and usefulness of a number of theoretical and empirical approaches to the understanding of the effect of technical change on the division of labour. But however briefly these ideas are discussed in the body of the text, it may still be useful to summarize some of the central arguments in yet briefer form in this introduction – in part to give readers a framework within which they may place the discussions which follow, but primarily in the hope that they may be tempted to read further.

The model of household choice between alternative technical means of provision for particular functions relies on a strategic simplification. It assumes that households face choices between pairs of alternative modes of provision; either they buy traditional final services, which require only that the purchasing household devotes some time to working for money (working in the 'formal economy') with which to buy these services; or else they buy various sorts of goods – domestic equipment and materials – which they use in a further production process outside the formal economy in order to satisfy an equivalent function. So this second alternative requires two different sorts of work; in the formal economy, in order to buy the goods, and in what we may term the 'informal economy', using the goods to provide services to the household.

Which alternative will the rational household choose? Within our simplified model, the choice depends on three factors – the household wage rate, the relative prices of final services and goods, and the amount of unpaid time necessary to use the goods to provide final service functions (which is, loosely, the 'productivity' of the goods). A limiting constraint on households arises from the fact that durable goods have to be combined with a certain amount of unpaid work time if they are to give rise to useful services. An hour's unpaid domestic work by a household member gives rise to a constant amount of services irrespective of that individual's wage rate in the formal economy. But the higher that individual's wage rate, the more he or she gives up by working for that hour in the domestic rather than the money economy. And the price of purchased services does not vary with the purchaser's wage. So, the higher the wage rate, the more sensible it is to buy final services rather than buying goods and forgoing paid employment in order to use them. It is frequently observed that richer households spend a larger proportion of their incomes on services than poorer households do (i.e. that demand for service is 'income elastic'); here we have an explanation for this phenomenon.

At any given wage rate, the higher the price of services relative to goods (and to the wage rate) the more sensible it is to buy goods rather than services. And the more productive the goods (i.e. the less unpaid work-time required to produce a given quantity of final service equivalents), again, the more sensible it becomes to buy goods. Over time goods have tended to get cheaper relative to services, because (for a number of reasons which we shall discuss later) manufacturing industry is able to achieve higher levels of productivity growth than final service industry. And technical change means that goods tend to get more efficient – more useful for the informal production of services, more 'productive' in our special sense – over time.

So the model gives us two predictions. At any point in time, better-off households will spend a larger proportion of their disposable income on services than worse-off. And over time, households at each particular income level will tend to decrease their proportion of income spent on services; this shift away from final service expenditure is an example of social innovation. These results are derived formally (Chapter 3) – and we are able to demonstrate that this process of substitution of goods for services has taken place quite generally across Europe (Chapter 6). The model is however obviously a simplification of reality. Through the book I shall attempt to make it a little more sophisticated. For example, I shall discuss the range of alternative social forms in which informal production occurs (Chapter 4); and I shall briefly consider the issue of the distribution of formal and informal work within the household (Chapter 9). But nevertheless the simple model is useful in that it provides a single basis for the understanding of a number of apparently disparate phenomena.

It gives us a prediction that demand for final services may sometimes decline with economic development, rather than increasing, as the conventional wisdom would have us believe (though this does not necessarily mean a decline in service employment – growth in intermediate demand for services and relatively low productivity growth in final service industries work against this possibility). Chapters 6 and 7 give empirical estimates of the relationship between social innovation and employment patterns in Europe over the last decade.

It gives us a basis for thinking about the evolution of unpaid working time (though Chapter 9 shows that the relationship between unpaid work and service consumption is by no means a straightforward one). It is also suggestive of prospects for the collectively provided public services; just as lack of productivity growth in marketed services has fuelled social innovation, so, perhaps, the same phenomenon in the non-marketed services has led in the past to 'tax revolts', and may lead in the future to pressures for new sorts of social innovations in the welfare state.

The social innovation approach may also give us a basis for solving two conundrums that have vexed historians and sociologists of science: the irregular relationship between inventions and the emergence of new markets for innovative products; and the irregular pace of economic development – the fact that some historical junctures seem to have greater potential for economic development than others. We use the concept of social innovation as the basis of a rather speculative explanation of these two problem issues (Chapter 5).

But the most important application of the model is much more practical. It appears that there may be an analogy between the wave of social innovations which dominated the developed world economies in the 1950s and 1960s, and a new wave, whose effects may reach us by the middle, and certainly will reach us by the end, of the 1980s. In the 1950s and 1960s a bundle of technologies (valves, and later semiconductors, fractional horsepower electric motors, the internal combustion engine), in combination with appropriate collectively provided infrastructure (roads, electricity generation networks), enabled a very wide range of social innovations in provision for domestic, entertainment, and transport functions. It may be that another bundle of technologies (cheap computing and data storage media) combined with appropriate (telecommunications) infrastructure, will enable another wave of social innovations in the 1980s – affecting the entertainment, transport, and communications, educational and even medical functions. What are the implications of the 'informatics revolution' for economic structure, for employment, and for work in general?

We cannot provide any very concrete answers to this question. The future social innovations we discuss in this book are, for the main part, still in the realms of science fiction – even though they, or innovations very like them, will be part of our everyday reality in ten or fifteen years' time. What we can do, however, is to provide grounds for a certain degree of optimism about our future prospects. The 'workplace centred' view of technical change focuses its attention principally on one particular sort of innovation, 'process innovation' – improvement in the efficiency of production of existing products for existing markets. Inevitably this leads to a certain pessimism about future prospects for work; we can make the same set of products more efficiently, it tells us, so we will need less workers. Our complementary 'household centred' view of technical change, however, enables us to tell the other side of the story – the promise of the new technologies for the development of new products, for new markets, and hence for new work opportunities, and new sorts of improvements in the wellbeing of our societies.

CHAPTER 2

A Straw Man: the March Through the Sectors

2.1 A Hierarchy of Needs

IN an attempt to reformulate a conventional wisdom, it is frequently convenient to establish a 'man of straw' – a superficially complete and convincing model which is nevertheless designed to come apart when attacked from the right angle. This mode of argument provides a straightforward framework for the writer; 'this is what you believe, is it not?' he says to his reader 'well you're wrong'. The cautious reader will quite properly be suspicious, and search for artificially weakened seams – the writer in turn is under a responsibility to ensure that his dummy argument is constructed from the most robust materials available. Readers of this present text should be reassured. The model of the succession of the sectors which follows is in fact quite a strong one. Though it will be put into question, it does not prove to be actually *wrong* – only *incomplete*. In this case, rather than knocking him down, we shall ultimately set the straw man back on his feet with some fresh stuffing and a new suit of clothes.

In 1887 the economist Robert Giffen addressed the British Association on the subject of 'The Recent Rate of Material Progress in England'.[1] His concern was the apparent reduction in the rate of increase of wealth over the previous decade. He presented time-series data on output of staple commodities – production of coal and pig iron, railway goods traffic, foreign shipping, consumption of tea and sugar – showing a considerable decline in their growth rate. He then considered a number of alternative explanations for this. These explanations (unsurprisingly, given the great age of the British economic disease) have a rather familiar ring: poor export performance and foreign protection, economically unjustified improvements in working conditions (shorter hours) promoted by the trades unions, bad weather, and the trade cycle, resource depletion, shortage of technical skills, even disillusionment with economic growth.

But Giffen, for a variety of reasons which need not concern us, rejects all of these as insufficient explanations. And, having found that he could not explain the phenomenon, he proceeded to question its existence. Had England's rate of increase of prosperity really declined? He cited three different social indicators to suggest that this decline had not in fact happened: the proportion of the population on the rolls of the poor-law administration had fallen faster in the decade in question (from 4.2 per cent to 3.0 per cent) than in the previous decade (where the decline was from 4.7 per cent to 4.2 per cent) from which Giffen infers increasingly full employment for the working classes; he cites an acceleration in the size of deposits (and steady increase in the number of depositors) in savings banks as evidence of improvement in the conditions of the lower middle classes; and while growth in income tax receipts had declined, the rate of increase of tax on income imputed from houses had been maintained; so, Giffen claimed, at least some part of the growth of prosperity of the tax-paying classes was continuing. This analysis posed a puzzle for Giffen. How could it be that, while there was a clear decline in the rate of growth of the staple industries, the increase of general welfare was continuing to grow unabated?

Giffen's explanation was that while the traditional staple industries were undergoing a relative decline, other new industries were growing. Most significant among these were what he called 'incorporeal functions' – we would now refer to them as service industries. He quoted data from the Population Census of 1871 and 1881 showing a quite striking growth in the proportion of the working population employed in 'transport . . . commercial . . . art and amusement . . . literature . . . education . . . '[2]

> We are apparently justified in saying that an increasing part of the population has been lately applied to the creation of incorporeal products. Their employment is industrial all the same. The products are consumed as they are produced, but the production is none the less real. If a nation chooses to produce more largely in this form as it becomes more prosperous so that there is less development than was formerly the case in what were known as staple industries, it need not be becoming poorer for that reason; all that is happening is that its wealth and income are taking a different shape.

This is, apparently, the first time that the notion of the service economy is broached in the professional economic literature. Whether Giffen was technically correct in his analysis of the British economy of the 1880s (he probably was not) is however not the point.

Simply, Giffen is associating a change in economic structure with a change in the nature of the society's marginal unsatisfied needs. In doing so he follows an important strand of nineteenth-century economic research – the analysis of household expenditure patterns. It had been demonstrated that household budgets reveal a quite clear regularity; the richer the household, the smaller the proportion of its income it spent on food, and the larger the proportion devoted to the satisfaction of more sophisticated needs. When he finds that over time an increasing proportion of what we would now think of as the national income is devoted to the 'incorporeal functions', therefore, he has a macro-economic manifestation of this pattern of household behaviour. The nation is becoming richer, more of the nation's households seek to satisfy the incorporeal functions, so the 'nation chooses to produce more largely in this form'. This line of argument has become the conventional wisdom of economics, so natural to us that we do not think to question it.

This conventional view of the process of change in economic structure relies on a generalization from comparisons of the life-styles of *different* social groups at *one* point in time, to the development of life-styles of particular groups at *successive* points in time. In the present lives of the rich, this view suggests, we see the future of the poor. This pattern of argument is implicit in much of our thinking about economic development; we move from 'cross-sectional' ('synchronic') views of the relationship between the patterns of consumption of the poor and the rich, to 'longitudinal' ('diachronic') projections of the change in the pattern of consumption as the society as a whole becomes richer. This is, I shall argue, a dangerously misleading approach to forecasting economic change.

This model starts with the psychological hypothesis of a hierarchy of common needs; first we need food, warmth, and physical security; once these needs are satisfied, then we become concerned about more abstract needs, for intellectual stimulation, for warmth and security in the society of others.

This psychological hierarchy translates itself – according to the model – into a hierarchical pattern of economic demand for commodities.

The traditional assumption – based ultimately on the findings of such nineteenth-century economists as Giffen and Engel – is that the very poorest households spend almost all of their budget on food, the slightly less poor spend a slightly smaller proportion on food, a larger proportion on durable goods, but hardly any more on services; and the rich spend very little of their marginal income on food and durables, but a very large and increasing proportion on services. Food, we assume, is an 'inferior' good, having a negative elasticity of demand with respect to income; services are a 'superior good' with a positive income elasticity of demand; durables are a superior good at the low end of the income scale, and an inferior good as we approach the top of the scale.

From here the traditional view moves to employment structures. When the majority of consumption is of food, clearly the majority of the population must be employed in food production. So, while the average household income is relatively low, the working population is distributed into the primary (agricultural and mining), secondary (manufacturing), and tertiary (services) sectors more or less in proportion to the average household's consumption of the products of the sectors; the majority of the working population is in the primary sector, with a smaller proportion in the secondary, and a smaller proportion still in the tertiary. As the society as a whole gets richer, household consumption patterns change so as to produce an employment profile with a majority in the secondary sector, a smaller and declining proportion in primary industry, and a smaller but growing proportion employed in service provision. The changes in the nature of the society's marginal needs produce the transformation from a predominantly agrarian society to a predominantly industrial society.

The next stage would appear to follow just as logically. As societies get richer still, their preferences for increases in consumption are progressively concentrated on services, so their employment profiles show small but stable agricultural workforces, declining numbers of manufacturing workers, and burgeoning service industries. This – though of course never so crudely stated – is the essence of the argument for the emergence of the Post Industrial 'Service Economy': as we get richer, our marginal needs change, this is reflected directly in a changing pattern of economic demand, which in turn is reflected in a changing pattern of employment. It should certainly not be suggested that economic growth has no impact on the nature of people's unmet needs; as societies get richer, of course, requirements at the margin of consumption will change. But, the *rest* of the argument is misleading. The means by which people's needs are satisfied themselves change, and this change – in the technology and social organization of final production – is largely absent from the conventional account.

2.2 The Productivity Gap

So far, however, we have only half the argument. In the modern version of the 'sectoral succession' model, the hierarchy of needs/demands is complemented by a second phenomenon – the 'productivity gap'. It is widely held that the service sector exhibits a lower rate of productivity growth than the rest of the economy. The modern conventional wisdom includes both elements; Kindleberger provides the classic description:

> The French statistician, J. Fourastié, who is much interested in productivity, has suggested a model of economic development which combines systematic differences in productivity by sectors with the pattern of income-elasticity of demand implied by Engel's Law. In his exposition, potatoes are chosen as an example of a primary good, bicycles as an example of secondary and a hotel room of tertiary output. Labour productivity in the first is said to have increased from 100 in 1800 to 130 in 1950. Consumption rose continuously throughout the period to the 1920s but has recently developed negative income-elasticity with per capita consumption 250 per cent of the 1800 level. In bicycles . . . productivity increased from 100 in 1900 to 700 in 1950, and is still increasing. Consumption has gone up nine times in the same period, but is beginning to level off. In the hotel room, labour productivity has remained practically unchanged from 1800 to 1950, but demand has brought about an increase in consumption from 100 to 1800, to 10,000 in 1950. Fourastié uses these systematic differences in productivity and demand by sectors to project systematic changes in the terms of trade which favour tertiary industry over primary, and both over secondary.[3]

This model was formulated, apparently independently, by a number of different economists in the late 1940s and early 1950s,[4] and it has proved remarkably durable; it stands, for example, quite unquestioned, at the heart of the standard sociological text on 'post industrial society'.[5]

Why should there be this productivity gap between manufacturing and service industries? Two different sorts of explanations are advanced. The first relates to the nature of the products. It is argued that services are inherently labour intensive, that the stuff of services is the face-to-face contract between the service producer and the service consumer. Since services are, in Adam Smith's phrase 'consumed as they are produced', the provider of the final service must be present at the instant of consumption. Furthermore, services are complex and are therefore difficult to automate. This position is not completely convincing; it becomes increasingly easy to think of technical means for getting round the face-to-face requirement, and also to enable a certain amount of automation in even the most complicated and personal of services. But nevertheless the argument has a certain plausiblity.

The second sort of explanation, however, is the more satisfactory – that the low productivity growth in services is related to the structure of service industries. Services seldom find themselves in a situation of anything approaching perfect competition; the number of service workers in a particular locality will be limited, and information about the quality of their production is hard to come by – for that matter, such information is difficult even to measure. In many service industries firms tend to be small, to use little capital equipment, and to employ low-paid, non-unionized labour. In others, the characteristic service institutions are large, bureaucratic, perhaps nationalized, and without any effective competition.[6] In general, therefore, we can conclude that pressures to promote productive efficiency are less in industries providing final services to consumers than elsewhere in the economy.

2.3 Social Innovation and the Succession of the Sectors

So we have what seems, on the face of it, to be a rather plausible model. On one hand, as societies get richer, the marginal unsatisfied needs of their populations change, from the satisfaction of relatively simple and basic desires to more sophisticated and complex ones. Accordingly their marginal unsatisfied demands change, from food to manufactured goods, and then from manufacturers to services. On the other hand, productivity rises slowly in the service sector, so that while technical change may lead to the displacement of manufacturing workers even while the output of the manufacturing sector is still growing, the same does not hold true for service industries – the number of service workers will always continue to rise.

What is wrong with this argument? The error resides in the direct identification of psychological needs with economic demands. This identification may be correct when we are considering cross-sectional evidence – after all, at a particular point in time a society may be expected to have a given set of means for satisfying needs (or, to use the vocabulary introduced in the previous chapter, a particular distribution of 'modes of provision of service functions'). So at a particular historical instant, the association of an increase in the proportion of income devoted to a particular class of commodity with increasing household wealth may well be a valid indication of the increasing marginal importance of a particular class of need. But, over time, the way needs are satisfied may change; the same need may, at different points in time, translate itself into demands for different sorts of commodities. At different historical junctures there may be different modes of provision for the same function.

In fact the 'succession of the sectors' model has conditions which lead to such changes in the mode of provision built into itself. The 'productivity gap', the gap between productivity growth rates in the manufacturing and service sectors, has the almost inevitable consequence that the prices of services rise relative to the prices of manufactured products.[7] So, where there are competing modes of provision for a particular service function, the mode that involves more purchased final services will be progressively disadvantaged relative to the mode that involves more purchased goods. Thus it may be true that at a particular historical juncture the rich had more servants, travelled more often by train, went more frequently to the theatre, than did the poor. But nevertheless, as the poor got richer over time, they did not employ more servants and buy more train and theatre tickets – instead they bought domestic machinery, cars, and television sets. And one of the crucial factors in these processes of social innovation was the changing relative prices of the two sorts of commodities employed by the competing modes of provision; the productivity gap may mean that particular services get priced out of their markets.

So the sectoral succession model contains an inherent contradiction. But this does not mean that we should abandon it altogether. In the following chapters we shall provide empirical evidence to demonstrate that social innovation means that sectors do not succeed each other in the manner suggested by the model, and in particular that marketed services have throughout Europe tended to decline as a proportion of consumer expenditure, being replaced in household budgets by the purchase of equivalent goods. But though expenditure classified by *commodity* (i.e. goods vs. services) does not conform to the model, expenditure classified by *function* (i.e. food, shelter, domestic, entertainment etc.) does so conform. The changing expenditure by function shows, as we shall see in Chapter 6, a very clear tendency to move from 'basic' to 'luxury' categories, confirming the basic prediction of 'Engel's Law'. So the model of structural change in the economy developed in this book does employ the notion of a 'hierarchy of needs' – *in combination with* the concept of social innovation.

NOTES

[1] R. Giffen, *Economic Enquiries*, 2, Vol. II.
[2] Ibid., pp. 134–40.
[3] The quotation comes from Kindleberger, C. P., *Economic Development*, McGraw Hill, 1958, his reference is to J. Fourastié, *La Productivité*, Presse Universitaire, 1952.
[4] As well as the French source cited, similar arguments can be found in, for example, C. Clark, *The Conditions of Economic Progress*, Macmillan, 1940.
[5] Daniel Bell, *The Coming of Post Industrial Society*, Heinemann, 1974, Ch. 2.
[6] These characteristics are discussed at length in J. I. Gershuny and I. D. Miles, *The New Service Economy*, Frances Pinter, 1983, Chapter 8.
[7] See ibid., Chapter 2.

CHAPTER 3

Patterns of Demand and the Division of Work Time

3.1 The Substitution of Goods for Services

THE view that the sectors of the economy develop in the same sequence as the pattern of final needs – from material to the non-material, from goods to services – is, therefore, too static. Purchase of commodities from the money economy may fulfil one of two functions. The transaction may itself be an indicator of an act of final consumption; when we pay for a cinema seat we are buying a service which we will almost immediately consume. But when we buy a television set, or a motor car, we acquire a good, which we will not consume, but *use* – probably in the production of a service. The purchase of a good can be regarded, not as final consumption, but as an intermediate activity, a *means* to an end which is usually the provision of a final service. If all purchases from the money economy were of genuinely final commodities – the things we actually need – then there would be very good grounds for identifying the sectoral development of the economy under conditions of economic growth with differences between the apparent needs of poor and rich people. And indeed some of the changes in economic structure are of this nature. The growth of the educational and medical sectors in the economies of all developed societies presumably reflects the structure of the marginal needs of their populations. But the purchase of final services from the money economy is not the only way to acquire them – the alternative is to buy intermediate commodities and investment goods and to use these in the production of the final services. This is the significance of technical change for the discussion of economic sectoral development: it can lead to a change in the way final production is organized. When we move from a comparison of the consumption patterns of the rich and the poor at one point in time, to changes in consumption over time as the society gets richer, we must also consider the possibility that the social organization of production will change – that there may be an alteration in the pattern of money economy demand for intermediate and final products associated with a particular category of human need.

3.2 Some UK examples

The changing pattern of final demand for marketed services in the UK is a good example of such a change. In 1959 services were clearly 'superior commodities' (Figure 3.1). The larger the family income, the larger the proportion of that income

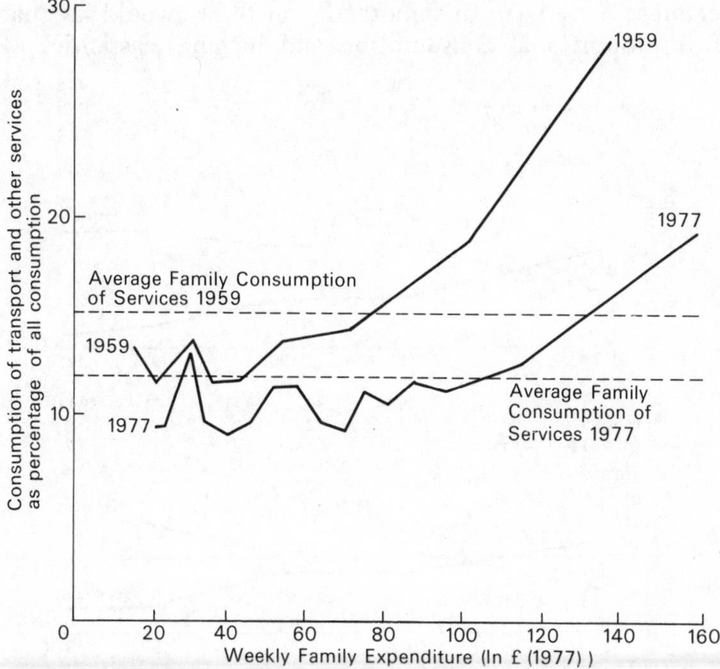

FIGURE 3.1 *Changing income elasticity of service consumption (household expenditure UK 1959–77)*

spent on services; and, with the exception of the very bottom of the income scale, the marginal demand for services with increasing income – the income elasticity of demand – itself increases throughout the income scale. The consumption of services in the population as a whole averages at a little more than 15 per cent of all household consumption[1] – but the richest group of families in the sample consumed nearly twice this proportion in the form of services. If we were to use this 1959 cross-section as a basis for a forecast of future patterns of consumption, then, assuming rising income, we would be likely to predict a considerable increase in the proportion of services to total consumption.

But in fact, the proportional consumption of these transport and other services has declined continuously since 1959. The average service consumption in 1959 was 15.2 per cent of the total, and by 1977 this proportion had fallen to 12.1 per cent. As we see from the 1977 cross-section, the society's marginal propensity to consume services with increasing income had declined; if we were to smooth the curve, we would say that both proportional consumption and income elasticities of

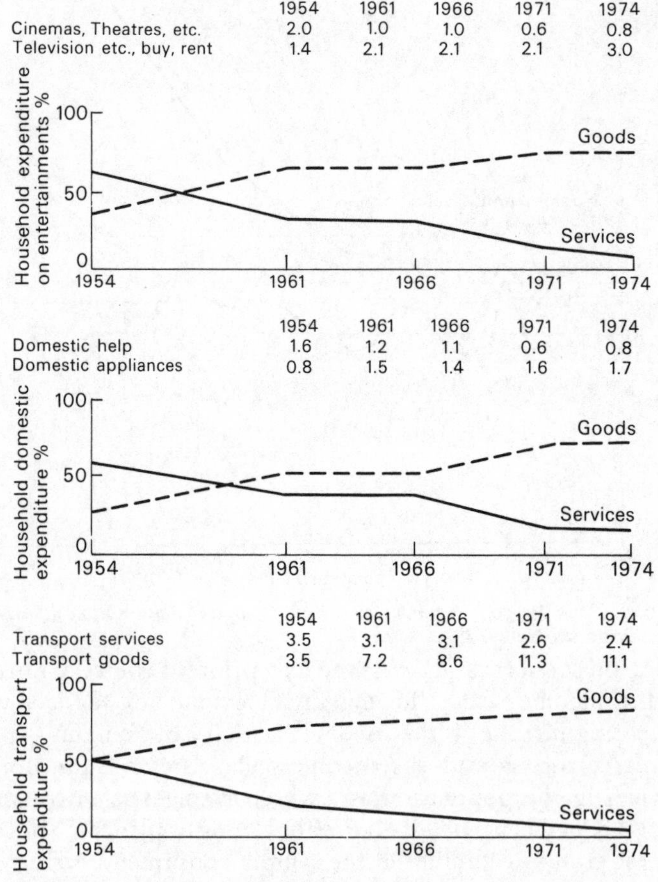

FIGURE 3.2 *Substitution of goods for services*

demand at each level of income were lower in 1977 than in 1959. Though there is still in general a positive income elasticity of demand for services in 1977 – services are still 'superior commodities' – the flattening of the cross-sectional curve outweighs the increase in incomes that took place over the period, so that service consumption is reduced.

We get a clue to the nature of this change when we look at substitutes for individual declining services (Figure 3.2). It appears that a process of substitution is indeed taking place – of domestic machinery for domestic services, of motor cars for purchased transport services, of television and stereo systems for cinema. Or more generally, a process of substitution of the purchase from the money economy of intermediate and investment goods where previously final services were purchased – a process in which the final production of services takes place outside the money economy. The consumption of service functions which, on the basis of cross-sectional comparisons, we expect to rise as a proportion of total household expenditure probably *is* rising, but the proportional increase does not appear in the accounts of the formal economy.

The reason for this change can be initially summarized in terms of changing relative prices in the formal economy. Prices of services have been rising at approximately twice the rate of

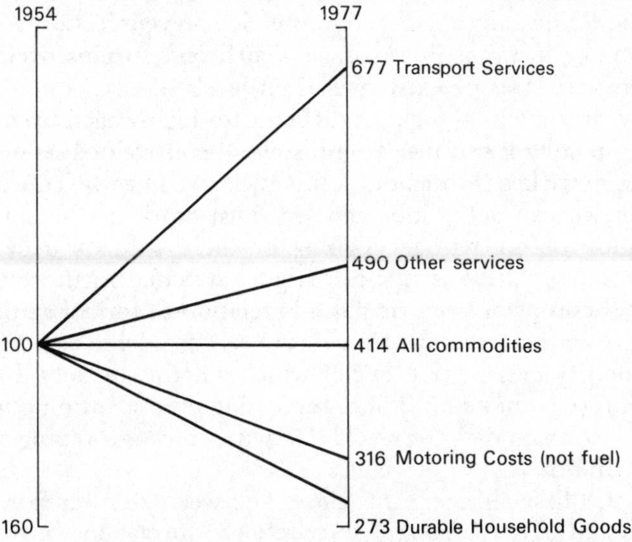

FIGURE 3.3 *Price changes UK, 1954–77*

increase of prices of goods (Figure 3.3). Simply: as goods get cheaper relative to services, it becomes increasingly *rational* for individuals to stop buying services, buy goods instead, and use the goods to produce services informally. But there is more to the change than just relative prices. The following section will discuss this process of 'rational choice' between the purchase of goods and services from the formal economy.

3.3 Time Budgets and Household Choice Between Goods and Services

What determines the changing balance between purchase of goods and services from the formal economy? To answer the question we have to introduce a notion of time scarcity. Activities are to some extent constrained by natural cycles, days and weeks and years and generations. Some activities are tied biologically to time cycles: we must eat and sleep and possibly recreate with a minimum frequency every day or every few days. Some cycles of activity are imposed by economic and social systems. Adults are constrained to work in the formal economy for some period during the day or the week. There is, of course, no reason in prinicple that we should not, for instance, work intensely in the formal economic system for one year, and not at all throughout the next – but there are economic and social sanctions which prevent it. Low wages may make it impossible to save a sufficient surplus over the current expenses of existence. High levels of taxation – tied to a yearly cycle – make it difficult for high-wage earners to build up such a surplus. There is also a constrained sequence among activities throughout a life cycle; we must be educated before we can get a job, and we must work in the formal economy before we are eligible for social security benefits. And, finally, there is a social stigma attached to those who evade these cycles – particularly in relation to formal employment. For these reasons time has to be considered as a scarce commodity even within cycles which amount to only a tiny fraction of a human lifetime. Increasing time spent on one of the constrained daily or weekly activities means forgoing time spent on others.

In the first chapter, the choice between the purchase of goods and of services, and thus between formal and informal systems of final provision, was described as being dependent on the relative prices and effectiveness of the two alternatives. We must clearly add to this a consideration of the opportunity cost of time spent in self-service activities. Increased work in the informal economy necessarily involves either less leisure time or less work in the formal economy. The inclusion of time into the calculation makes the picture slightly more complicated.[2]

We cannot expect to construct an adequate general explanation of the choice between goods and services which includes all the motivations which underlie the choice, but we may gain some insight from a much more limited model that considers just the constraints of economic rationality. We shall assume that the household is simply motivated to maximize its final consumption of services, and that it is absolutely indifferent between work in the formal and informal economies as a means to achieving that maximization. The rational household may be expected to balance the time it spends working in the formal economy against the time it needs to use whatever goods it buys for the informal process of production of services.

If it works long hours in the formal economy, then it earns a lot of money, but it has only a little time left for informal production activities. Since it wants to maximize its total consumption of final services, it will be likely to spend a large proportion of its earnings on services produced in the formal sector, and only a small proportion on goods to be used in the informal production of services – it will buy services rather than producing its own. If, however, it works relatively short hours in the formal economy, then it will have a lot of time left over for informal production activities, and it will be likely to spend a large proportion of its money income from the formal economy on goods, which it will use in the 'direct production' of services for itself.

We can make this picture rather more determinate. Assume further that some informal activities are more productive than others. Our household will clearly wish to spend its time in the more productive rather than the less productive activities. This suggests a criterion for choice between formal and informal production activities. Stated approximately (we will describe it precisely in the following section) it is as follows: where the price of a service purchased from the formal economy is less than the money earnings the household forgoes in order to spend its time producing that service for itself plus the cost of the goods and materials which would be used in this informal production, the household's *rational* course of action – within our assumptions – is to purchase that service, and to work the extra period in the formal economy to pay for it. If the household's wage rate is high relative to the cost of services (that is principally the wage level in the consumer services sector), then it will work long hours in paid employment and *buy* services, whereas if its wage level is relatively low it will buy goods and produce its own services.

The wage rate is not the only variable, however. Over time, as a result of technical change and investment in producers capital, the price of consumer goods declines and their productive potential increases. So holding relative wages constant our household can afford more productive capital as time passes, and hence the terms of trade between the formal and the informal economics change in favour of the informal economy. Time spent in informal production becomes more productive; for any given task, a given time spent in self-service activities would have to be matched by an increasing purchase of the equivalent service from the formal economy in order to justify its purchase. We would expect, therefore, that the decreasing price of goods relative to services, and their increasing production potential, would lead to a transfer from formal to informal production.

We can put this in a more concrete form. A clerk in 1900 would, on the basis of a rational calculation, work all day at his profession and pay a servant to work all day at dusting and sweeping. The clerk's average wages was many times that of the domestic servant while his own productive capacity for domestic services was presumably no higher than that of the servant; it, therefore, paid him to pay the servant. The technical innovation of the electric vacuum cleaner, however, increases the clerk's domestic productivity. At the same time the wages of the domestic servants have risen somewhat. So in 1978 the rational clerk will, all things being equal, rather than employ even a part-time servant, work a few minutes less each day and clean his own house.[3]

There are two different processes here. The first might be considered a *relative wage* effect; at any point in time, or with given relative prices and productivity levels, the higher the wage the more services a rational household might be expected to buy, and the longer the hours of work in the formal sector as against informal production activities. The second is a *relative price* effect. As the price of goods declines against the price of services over time, rational individuals will tend to transfer gradually from the purchase of services, and hence employment in the formal economy, to the purchase of goods, and work in the informal economy.

These two effects, though analytically distinct, cannot in practice be separated. As time passes, real income rises – or at least has in the past risen – in developed economies. This must (at least under conditions of full employment – we will return to this proviso later) lead to a change in *relative* incomes within the economy. The formal provision of final services is necessarily labour intensive and is not subject to manpower productivity improvements, whereas the production of goods is subject to such improvements. Or, to put it less strongly, productivity growth is certainly slower in the services than in manufacturing. The effect of real economic growth, therefore, goes in one of two directions. Either the price of services (and hence the level of service wages) rises, less services are consumed, and total service employment declines; or services wages are forced down to compete with goods, and service consumption is maintained or increased. The latter possibility is, however, unlikely, at least in a buoyant economy; rather than accept lower wages, service workers would be likely to seek jobs elsewhere in the economy. The former direction of change is, therefore, to be expected – and was indeed the case in the expanding economies of the 1950s, 1960s and 1970s; wherever there was technology available to encourage the informal production of services – notably in transport, and entertainment, and domestic services – the formal provision of services declined.

We should immediately say that this argument ignores some very important institutional constraints. Individuals are not usually free to vary their hours of paid employment – which might make adjustment in the distribution of working hours rather difficult. But remember that we are discussing, not individual but *household* behaviour; some households do very clearly make just such adjustments as we are describing

– between the wife's full-time employment and various degrees of part-time employment. Which brings us to the most important qualification: it would be possible to use the model developed here in an attempt to explain the division of labour within the household, using the difference between male and female wage rates as a justification of the disproportionate female responsibility for domestic work – this is no part of my present intentions. Such analyses are subject to the criticism that they reverse cause and effect (i.e. the wage differential is a consequence of the household division of labour rather than vice versa) – and in any case our everyday experience tells us that the division of labour within our households relates, in general, more to sociological phenomena (i.e. our role-models) than to economic.

The model which I have developed, and which will be further elaborated in the following section, should however be interpreted within a rather more general framework. It is developed here in order to give some more definition to our view of how households go about choosing between alternative modes of provision of service functions. Its interpretation will be limited, in this book, to a view of what underlies households' strategies of allocation of work time and expenditure over the long term. (The next section contains a formal presentation of the model; those readers who dislike geometric or algebraic formulations of economic arguments may proceed to section 3.5 without missing a very great deal.)

3.4 A Model of Choice Between Formal and Informal Production of Services

We start with a conventional budget curve analysis: given a particular structure of preference between goods and services, the proportion of the consumer's total expenditure devoted to each of the two commodities is dependent on the slope of the budget curve, which is in turn dependent on the prices of the commodities. Holding preferences constant, the higher the price of one commodity, the larger the proportion of expenditure devoted to the other.

Next, let us assume that people divide up their time as follows:

A.1 – first they decide on what proportion of the day

they wish to work, irrespective of whether that work be for money or in unpaid household tasks

A.2 – they then decide, on the grounds described below, on what proportion of that total working time is to be spent in paid work and what proportion in unpaid.

This is a rather strong assumption – and, on the face of it, a rather unlikely one. But it is a heuristically useful one, representing as it does the choice we all face between buying final services ready-made or providing our own; the real-life condition is perhaps weaker than the one we assume for the model but it is nevertheless present. (In Chapters 8 and 9 we shall consider some of the empirical evidence about households' decisions about total work time, about its distribution between formal and informal production – and about its distribution among household members.)

We must assume additionally

B.1 – that the use of consumer goods takes some finite time – the more goods, the more domestic production and the more time required (– and that service consumption takes no work time)

B.2 – that the result of domestic production is 'informal services' which substitute for services that might otherwise have been purchased from the formal economy

B.3 – the reason for buying goods is to produce such 'informal services' (i.e. goods are intermediate materials for informal production).

Assumptions A and B together impose a constraint on the conventional budget curve – given the limited total work time, the more devoted to work for money, the less time is available for domestic work – hence the relationship found in Figure 3.4a.

Between budget curves b_1 and b_2, the hours of work are short enough that even if the entire budget were spent on goods, there would still be enough time to use the goods for domestic production. But as hours of work increase past that for b_2, the situation changes: the more time spent working for money, the less is available for domestic work. So, a household which works sufficient hours to earn budget b_4, for example, finds itself in the situation that if it buys *more goods than that represented by point* p_4, it has insufficient time left to use them.

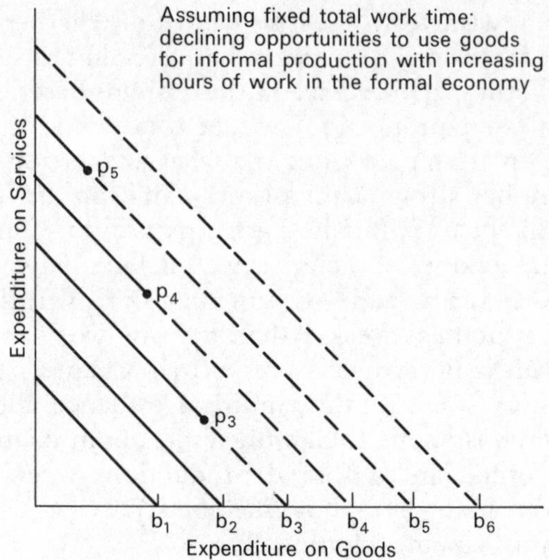

FIGURE 3.4a *The time constraint*

Let us consider this household earning enough to afford b_4:

> if it purchases *more* than p_4 goods, then it is *acquiring* goods that it has insufficient informal work time to use properly and forgoing thereby services that it could consume – it reduces its welfare.

Assume further that a given marginal expenditure on goods generates a larger quality of final, informal services than could be purchased from the formal economy. (This is a reasonable assumption since £1 spent on goods buys informal services worth £1 + the value of whatever household labour and overheads are input to the process of informal production of final services.) So:

> if it purchases *less* than p_4 goods, then it is *forgoing* a given quantity of informally produced services in exchange for a lesser quantity of formally produced services – again it is failing to optimize its welfare.

(Of course it may choose to work shorter hours – accept less formal services in exchange for more leisure time – but this runs contrary to the assumption in A above, so we will ignore this possibility for the moment.)

This implies that, under the conditions stated, the actual range of choice for a welfare maximizing household is the locus of such points as p_4, p_5, etc. – which we may refer to as the 'time budget curve'. Each point on the time budget curve represents a particular combination of purchase of goods, purchase of services, time spent in formal work and time spent in informal work. And, as with the conventional budget curve, the proportion of goods to services purchased must be dependent on the slope of this curve: the steeper the slope, the more services are purchased.

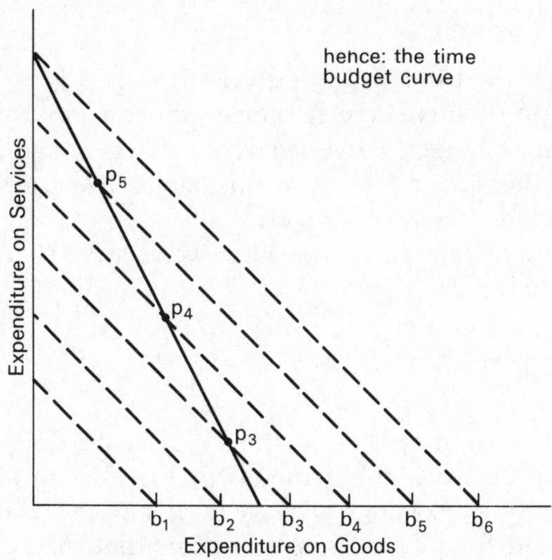

FIGURE 3.4b *The time budget curve*

We can derive the time budget curve algebraically as follows: The conventional budget curve has the form:

1. $S = \frac{1}{s} I - \frac{g}{s} G$ where G = quantity of goods
g = price of goods
I = income
S = quantity of services
s = price of services.

Income is the multiple of the individual's wage rate and his work time in the formal economy, so, (embodying our assumption of a fixed total amount of time reserved for work),

2. $I = w(T - T_i)$ where w = wage rate
T = total time available for work
T_i = time spent in informal work.

And we can for convenience assume that the relationship between the quantity of goods and the time taken to use them is a simple linear one:

3. $T_i = pG$ where p = coefficient of combination of goods with informal working time.

Solving these three equations we arrive at:

4. $$S = \frac{wp}{s}T - \frac{wp + g}{s}G$$

. . . which is the time budget curve.

The slope of this curve is the proportion $(wp + g)/s$; as in conventional budget curve analysis, if we hold preferences constant, the larger its value, the larger the proportion of expenditure devoted to services.

So holding preferences for all work relative to leisure constant, we arrive at:

5. $$S = f\,\frac{wp + g}{s}$$

This equation summarises, in the simplest possible form, the relationship between the demand for formally produced (i.e. purchased) final services, and wage rates, and the relative prices of goods and final services. We shall see that this relationship goes some way towards explaining both the cross-sectional and the longitudinal patterns of demand for services exhibited in figure 3.1.

3.5 Time Budgets and the Demand for Services

It should be stressed that this argument cannot be considered as providing in any way an adequate behavioural explanation for the allocation of time to the alternatives of paid and unpaid work. Particular individuals' actual behaviour will be determined by a range of factors which are not included in the model: the individuals' personal history – including the behaviour patterns of their parents or other formative influences; the norms of the culture within which they are living; their stage in the family life-cycle (particularly whether they have children); the geography of their immediate neighbourhood (i.e. the accessibility of schools, workplaces, and recreational facilities); and the number and type of people in their households.

So what does the model tell us? It identifies a pattern of behaviour which yields, in some sense, objectively optimal results in terms of the ultimate level of services received by the household. How this pattern of behaviour is actually acquired, the set of processes whereby the life-style actually evolves, is not described by the model. In all probability, it does not develop by any ratiocinative process that resembles a comprehensive rational evaluation of available alternative activity patterns; more likely, it develops through an incremental series of accidental happy discoveries and the emulation of attractive examples. In reality, we might expect households to encounter new ways of organizing their time more or less at random, and that the more beneficial options they encounter will tend to persist, replacing less beneficial ones. The model identifies a set of constraints which determine the objectively most beneficial pattern of behaviour; actual behaviour patterns will tend to evolve in the direction of this ideal – but the model does not in any way describe how this evolution occurs.

We have used the model to demonstrate that the optimal proportion of households' expenditure devoted to services may be considered to be a function of the following fraction:

$$\frac{(\text{Household wage rate} \times \text{Time needed to use goods}) + \text{Price of goods}}{\text{Price of services.}}$$

From this fraction we can derive the various effects that we describe in Section 3.3. The higher the household wage rate – holding the other variables constant – the larger the fraction, and hence the larger the proportion of household income devoted to services: this is what we previously called the 'income effect'.

Holding the wage and time variables constant, the higher the price of services relative to goods, the smaller the fraction, and hence the smaller the proportion of household income devoted to services; this is what we called the 'price effect'. And we can also derive a domestic productivity effect; the less time needed to use equipment to produce given output (i.e. the more 'productive' it is), the smaller the fraction, and the smaller the proportion of household income devoted to services.

We have good reasons for expecting the prices of services to rise relative to goods over time. There is the 'productivity gap' previously mentioned; assuming that service workers wish to maintain their wages relative to equivalent employees in manufacturing industry (and that the 'labour share' of value added does not decrease), the price of service must necessarily rise. Fig. 3.5 shows the time budget model's prediction of the relationship between the proportion of household income devoted to services and household wage rates – at two different points in time, where the relative price of services is increasing over time. It may be seen as a reasonable first order approximation to the pair of curves found in figure 3.1. Our time budget arguments in short, may be seen as a micro-economic model of the process of social innovation, whose macro-economic consequences are seen in the declining income elasticity of demand for services over time.

This result is of great importance. Time budgeting models of the sort developed by Becker, Linder, and Gronau have

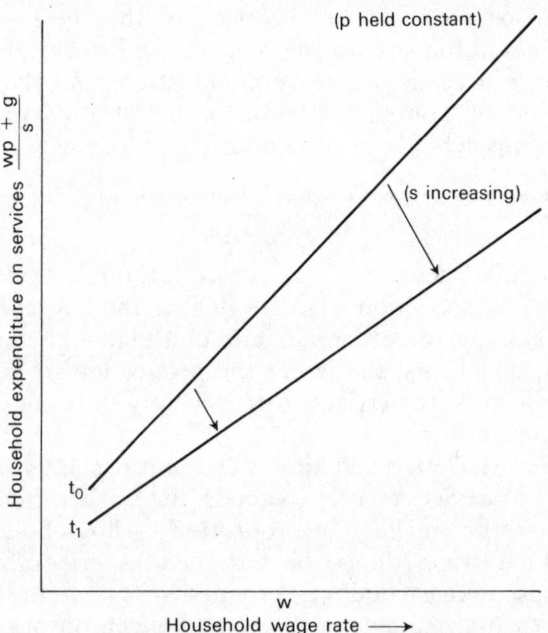

FIGURE 3.5 *Time budget predictions of wage rate elasticities of demand for leisure patterns – the sorts of concerns covered in Chapter 8, services.*

typically been used to explain issues of labour supply and leisure patterns – the sorts of concerns covered in Chapter 8, and more particularly Chapter 9 of this book. Here, however, the model is being used for a quite different purpose: the investigation of the determinants of demand for commodities – the balance between demand for goods and for services. Figure 3.5 provides the bones of an explanation, both for the positive income elasticity of demand for services, which gives us the 'Engel's Law' effects, and for the negative price elasticity of demand for services – which underlies social innovation.

NOTES

[1] Comparing consumption between these widely separated dates is complicated by the very different price trends exhibited by different commodities (see Fig. 3.2). If we were simply to inflate service expenditure in 1959 by the Retail Price Index in order to compare it with 1977, we would underestimate 1959 consumption since the preives of services have risen faster than the average. To take account of this differential price rise, service expenditures have been inflated by their individual price indices. If this procedure had not been adopted, the average service consumption by households in 1959 would have been calculated as 12.7 per cent and the 1959 curve in Fig. 3.1 would be slightly flattened, but the declining income elasticity of demand for services between the two periods would still be clearly visible.

[2] The model proposed in the following pages bears some formal similarity to the work of such economists as Becker, Linder and Gronau (Becker, G. 'A Theory of the Allocation of Time', *The Economic Journal*, Vol. XXX No. 200, 1965, pp. 493–517; Gronau, R. 'Leisure, Home Production and Work; The Theory of Time Revisited', *Journal of Political Economy*, Vol. 85 No. 4, 1977, pp. 1095–1124; Linder, S. *The Harried Leisure Class*, Columbia University Press, New York, 1971.)

[3] Of course, the vacuum cleaner also potentially makes the servant more productive: the argument in this case reduces to the proposition that the organizational and motivational advantages of self-servicing will mean that its effective increase in productivity will tend to exceed increases in productivity growth in purchased services.

CHAPTER 4

Informal Production Systems

4.1 Work Outside Employment

WE might pause, at this point in the exposition, to discuss some of the wider implications of the argument so far. The concept of social innovation has led us to consider an area of productive activity which we must certainly think of as 'work' – but which nevertheless falls quite outside the conventional economic definition of 'the economy'. In the next chapter we shall resume our discussion of the consequences of social innovation for employment in the money economy; here we shall discuss its implications for work in the 'informal economy' (which is of course not really an economy in itself, but rather the neglected part of the full economic system which includes both formal and informal production).

Our discussions so far have led us to think about the production of services within the household, using unpaid labour and domestic capital equipment, which is progressively substituted for the production of services in the formal economy. As we develop this trend of thought, we will find that non-registered economic activity also takes place in locations other than just the household – in communal groups, and in the 'underground' or 'black' economy. And it will become clear that the rationale for the development of these sectors of production bears much in common with the model of social innovation described in the previous chapter. In the following sections we shall look briefly at the nature of these alternative economic sectors, at the implications of recognizing these sorts of production for our view of the process of economic development, and at how we might seek to measure them. And, finally, we shall consider why these developing areas of production raise worrying problems for social policy.

4.2 The 'Circular Flow of Income' and Informal Production

The formal economy can be conceptualized as consisting of

flows of money and commodities between households and the formal production system (Figure 4.1a).

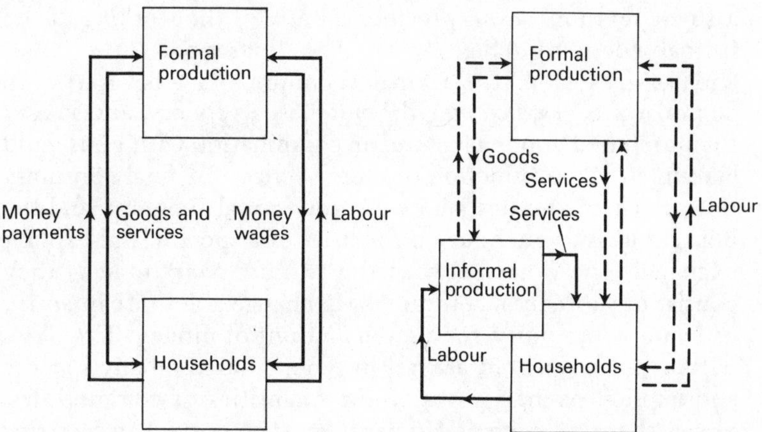

FIGURE 4.1a *The formal economy* FIGURE 4.1b *The informal economy: household production*

Households put labour into the formal production system, and receive in exchange money wages. They buy goods and services from the formal production system in exchange for money payments. These exchanges are *specific*: each commodity flow, be it labour, good or service, in one direction is balanced by a money flow in the opposite direction. The sum total of the money flows in one direction – either total wages or total payments for goods and services – may serve as an indicator of the extent of formal economic activity. Each of these flows is accounted for by some arm of government or by some official survey. In the UK, for example, the tax and National Insurance authorities, together with the annual Family Expenditure Survey, Census of Production, and the Census of Employment, can in principle ennumerate all of these flows – and this ennumeration forms the basis of the national accounts.

It has always been recognized that there is some economic activity *outside* these flows. Subsistence agriculture in developing economies, for example, would certainly not appear in this sort of accounting framework. And in developed economies, there is ready recognition of the statistical anomalies that result from such events as the marriage of a householder to his or her erstwhile housekeeper. Nevertheless, the nature of these sorts of unaccounted economic activities is rather different from the accounted ones. We can symbolize this by a slight revision of our previous picture of the working of the formal economy (Fig. 4.1b). The household buys goods and services from the formal economy. As I have tried to demonstrate, it *consumes* the purchased services, and it *uses* the purchased goods *as capital*, in combination with household labour, in the production of more services for final consumption. Each of the flows between the formal economy and the household is carried out on the basis of a specific and explicit exchange; the flows between the two are *conditional* – they consist of short-term contracts to exchange a definite quantity of some commodity for a given amount of money. The flows within the household are not in general based on any specific and explicit exchange of definite quantities of commodities. Social theorists such as Homans[1] may be correct in asserting that in the longer term there may be exchange relationships implicit within family structures, but these are rarely specific to particular historical instances – that is, a wife can rarely be said to have carried out some particular household operation on some particular day in exchange for some other particular household operation carried out by her husband. And these exchanges are typically between non-quantifiable or incommensurable values – as for example the exchange between cooking and social status. So we could distinguish (1) exchange between the formal production system and household, which are 'specific' – explicit, quantified, relatively short-term, and (2) exchanges within the household, which are 'generalized' – implicit, non-quantified and often very long-term – indeed often never consummated.

Closely associated with the household production system, and encouraged by the same social and technical developments, is what might be called the 'communal' production system. Included in this sector are 'voluntary' or religious organizations, baby-sitting circles, transport co-operatives, housing improvement co-operatives. At one extreme are those organizations on the verge of the formal production system – the baby-sitting circles or car pools – which operate on the basis of a quasi-money exchange, tokens or credits, and which break down if equal values are not exchanged within a relatively short period.

But more generally there is only a rudimentary system of specific exchange, and the major reason for the involvement of those who carry the burdens are the intangible, symbolic, unquantified, returns for their activities. What distinguishes this category of production is that real money is not used as an indicator of exchange of value for value. Where money is paid, in this sort of system, it is explicitly not in exchange for value received – so officers of communal organizations are paid 'honoraria', in *recognition* not *exchange* for services, and 'expenses', even though they may be fiddled, are rarely used as complete compensation. And in extreme cases – such as making meals for sick neighbours – the form of exchange can hardly be differentiated from that within the household.

As the previous chapter suggested, growth in the household economy (and in the closely associated communal sector) is promoted by the rising cost of purchased services relative to the declining cost of 'domestic capital goods'. Jobs are exported from the formal economy into these sectors of the informal economy; and as it becomes economically rational for individuals to provide services informally, it becomes correspondingly difficult to persuade people to purchase formally provided private services, or to pay higher taxes in order to receive more public services. At the same time, jobs are being lost in the manufacturing sector; so we expect rising unemployment. (Of course we may also expect unemployment for reasons other than social innovation.)

In a theoretical 'free' labour market, this situation would be self-correcting. Wage levels would fall, particularly in some relatively low-skilled service occupations, as a result of the entry into the market of labour displaced from manufacturing industry, with the dual effects of lowering the price of services, and thus increasing the demand for them, and at the same time (because of the lower price of labour relative to capital) increasing the labour input per unit of service output. By this argument, wages would fall until all those looking for jobs were employed.

But in most developed economies, labour markets are not 'free' to behave in this manner. There are inflexibilities in the formal economy's demand for labour resulting from employment protection legislation, employers' social security contributions and labour union restrictive practices. And the supply of unskilled labour is restricted by the high effective marginal tax rates which result from the high rate of loss of social security benefits with low earnings. These market 'imperfections' mean that full employment in the formal economy normally cannot be achieved by lowering wage rates.

There is however another sector of the informal economy where these 'imperfections' are not present: the 'underground', 'hidden', or 'black' economy. This third category of informal production activity is very close to the formal system. In fact it exists in the interstices of the formal economy, consisting largely of economic activities also undertaken in the formal economy, often by the same people. It is distinct from the formal production system, in principle, because those involved in it wish it to be so; it consists of economic activities which are hidden from the state authorities because of their illegality – either through avoidance of tax or other regulations, or because they involve thefts. Activities in this sector also bear some similarity to those in the other two informal sectors since, though the main mode of underground sector exchange is specific and money based, there is still a strong subsidiary element of generalized exchange. So, for example, Mars and Henry find a wide range of symbolic and unquantified values in people's explanations of why they are involved in occupational theft – 'helping out a friend' or 'because it's exciting'.[2] Even though these explanations may be coloured by a need to justify illegality, they do presumably reflect some reality.

The underground economy, which is by definition free of external restrictions, may to some extent counteract the inflexibility of the formal labour market; wages are not free to fall in the formal economy, but a low-wage informal sector develops instead. There are also reasons other than rising unemployment for the growth of this sort of economic activity. The falling cost of capital makes it continuously easier for individuals to own their own tools – which makes 'own account' working more convenient and less easy to detect. And the high costs of working within the formal sector, both high marginal rates of taxation and the administrative costs of dealing with the official state, make it attractive for those active within the formal economy to transfer all, or more usually part, of their activities into the underground economy.

Underground production activities may be classified into three groups. The first category ('type A'), consisting of the diversion of products from intermediate to final production, is always associated with formal employment – and though these activities are not included in estimates of the National Product, nevertheless they leave their trace in national accounts as costs absorbed within final production. Occupational theft is conveniently accounted as a producers' cost – but the stolen products must go somewhere, and indeed constitute part of someone's final consumption. Such theft is one extreme of a continuum passing through tax evasion to, at the other extreme, tax avoidance. The main distinguishing feature between the various points on this continuum is the connivance of the employer. Occupational theft is in the main conducted without the co-operation of employers, although apparently some do tacitly accept the right of employees to supplement their wages by 'fiddling' (Mars and Henry instance restaurants here). Occupational theft with the explicit co-operation of the employer is a frequent form of tax evasion – a senior executive with an expense account whose outgoings are never inspected by his company accountant, falls into this category. And some legally sanctioned business expenses – car and clothing allowances, for example, which are largely included in the national accounts as production costs, should appear as final consumption.

The other two sorts of underground production are less specifically related to formal employment. A second distinct category ('type B') is constituted by employment in a manufacturing process that feeds into the normal production system but is nevertheless not part of it. Outworkers 'making up' in the clothing industry, whose employment is disguised in the books of conventional firms, are a good example here. They are paid low 'piece-work' rates, they work at home, enjoy no protection from trade unions or public 'health and safety at work' legislation, they pay no taxes, and their employers pay no employment taxes. The third category ('type C') of underground employment is in the production of final goods and services directly for consumers.[3] In the UK this form of production is most clearly visible in house repairs, renovations and secondary construction, but elsewhere there are reports of its significance in professional services (Germany and the USA) and in manufacturing (Italy).

4.3 The Process of Development

Economic development is normally viewed as a one-way progress; a march through the sectors, from reliance on primary production through manufacturing production to a society whose major efforts are devoted to the production of services; a transition from a traditional society in which economic relationships are based on custom, in which such processes of exchange that exist are 'generalized', to a modern society in which an increasing proportion of social relationships are 'monetized', converted from 'generalized', to 'specific' exchange. The argument this book suggests that this view is wholly misleading.

Instead, it may be proposed that technical innovations, changes in capital endowments, modifications in legal institutions and in patterns of social organization, combine to produce a rather less tidy pattern of development. Rather than the steady one-way flow of economic activity from the household or communal basis to the industrial production system, Polanyi's 'great transformation', we have to consider a whole series of little transformations of production – 'social innovation' – perhaps taking place simultaneously, between the formal economy, the household or communal sector, and the underground sector; transformations in the mode of provision of final service functions, whose directions are determined by the particular social and technical conditions pertaining at particular points in time.[4]

We can list some of the circumstances which bring about these transformations. Consider the twelve possible transformations of production among the formal economy, the

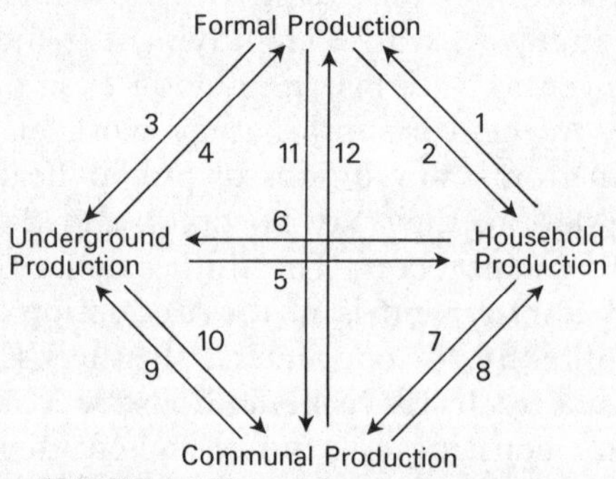

FIGURE 4.2 *The twelve transformations*

household, the communal sector and underground production (Fig. 4.2).

Transformation 1, from household/communal to formal production, is the conventional development process, relying on the economic principles identified by Adam Smith – economies of scale, division of labour, advantages of specialization, efficient use of factors of production. The processes he described are of course still continuing. And indeed, in the production of a wide range of commodities, these processes are still the predominant determinants of the pattern of development – but not for all commodities. Social and technical conditions have changed. In Chapter 3 it was argued that the change in the relative prices of domestic capital goods and service labour have led in the past to a transfer of production of some services from the formal economy into the household, and that this process is likely to extend to other services in the future: thus, transformation 2. We cannot, at this point, specify complete conditions under which each of these transformations are likely to take place, but we can suggest that this sort of transformation is likely where the capital goods used in the production of a particular commodity become so cheap that they need not be used intensively (and where the necessary skills may be expected within households).

Three conditions for transformation 3 have already been mentioned: unemployment combined with an inflexible formal labour market, declining real price of capital, and high costs in formal production. It may be useful to add a fourth, the criminalization of particular sorts of formal production; so, for example, a government attempting to restrict the use of labour-saving capital equipment in the formal economy, might simply drive such production underground. Transformation 4, from underground to formal production, has the converse conditions. Full employment might tempt workers from insecure underground employment to better protected formal jobs. Lower tax rates would reduce the attraction of the underground sector. And more extensive or effective policing would certainly raise the costs of involvement in underground production.

Examples of transformation 5 from underground to household production are rather difficult to find at present. We might perhaps speculate as to the effects of the decriminalization (in 1961) of the unlicensed brewing of beer for domestic consumption. Or more generally, we might imagine the effects of the introduction of construction materials which are less demanding of skilled labour (as for example plastic hot water pipes, currently uncommon in the UK), in reducing 'underground' construction activities in favour of 'do-it-yourself' household construction. There are close analogies here with transformation 2. Transformation 6 from household to underground production has converse conditions to those of transformation 5. Here we can find an example of criminalization of a household activity dating before the industrial revolution; certain forest rights possessed by the community before enclosures were revoked by Enclosure Acts, so that those who continued to trap small game were subsequently classed as poachers. Of more present day relevance, the existence of high unemployment may encourage the growth of low paid 'hidden' domestic employment – a transfer of production from the household to the underground economy.

An example of transformation 7 might be a change from the care of small children within the household to communal crèche facilities run by voluntary labour; category 8 might be the care of aged relatives, once the responsibility of a wider social group (the extended family), which may now fall entirely to one household. Transformations 9 and 10 might represent the alternative strategies of either having a household's children cared for in a communal crèche, which carries with it the responsibility of devoting unpaid labour time to support the crèche, or evading this responsibility without incurring the heavy costs of purchasing formal economy child care services, by leaving the children in the care of an unregistered child-minder.

The area covered by transformations 11 and 12 may turn out to be of increasing importance over the coming years. Most publicly provided services are very labour-intensive – perhaps 70–80 per cent of all the costs of public services are the salaries of public employees. The state now finds itself under two conflicting pressures: on the one hand, demands for extensions in public services; on the other, demands for reduction, or at least a secession of growth, in public expenditure – presumably as a result of public dissatisfaction with the effectiveness or efficiency of public provisions. The communal sector provides an opportunity for the resolution of this conflict – another sort of social innovation which increases the efficiency, or at least lowers the money cost, of the provision of a particular final service function. Rather than providing completed final services, the state might – as for example in the care of very young and very old people – provide the material equipment and infrastructure, building and furniture, books and toys, and medical equipment, together with 'intermediate services' in the form of professional advice, which would then be used by community groups to provide the final services themselves, using their own direct, unpaid labour.

The argument is simply that, at any point in time, the particular circumstances of technology, labour supply, and public regulation and organization, may lead to any of a wide range of different modes of provision of final service functions. Certainly, in the two centuries since *The Wealth of Nations* was published, the *aggregate* effect of all these 'little transformations' has been the 'great transformation' from household/communal production to formal industrial production. But over the last few decades the transformation of some sectors of production from a formal economic basis to a household basis has been of critical importance to the development of the economy as a whole. And it could be that in the future, the aggregate effects of the transformations 2 and 3 and 11 may outweigh the effect of transformation 1.

4.4 Measuring Informal Production

The argument so far is no more than theory. Certainly, the previous and following chapters give reasonably substantial evidence of declining proportional formal expenditure on services – but the growth of household production of services is merely *inferred* from the growth of household ownership of domestic capital goods. The growth of underground production of many different classes of commodity is now frequently reported in newspapers – but this is no more than hearsay. Can we expect in any reliable way to measure the informal economy?

Let us start with the underground sector. There have been two very ambitious attempts to measure the US underground economy, using aggregate money supply statistics.[5] The more convincing of these relies on the observation that underground economic transactions, in order to stay hidden, must depend on cash payments; thus, the author (Gutmann) assumes, any expansion of demand for cash as opposed to bank deposits is likely to indicate the expansion of the underground economy. The author observes that the ratio of cash to bank demand deposits in the USA declined until the period of the Second World War, at which time it started to grow again; he sees in this reversal the influence of the black market that developed in response to wartime shortages. Gutmann takes the average ratio of cash to demand deposits between 1937 and 1941 as the proportion of the money supply necessary for legitimate business, and assumes that any increase in the cash supply above this level is used in the underground economy. Assuming further that the velocity of circulation of the cash supply is the same in the underground economy as in the formal economy, he arrives at an estimate of underground production as constituting 10 per cent of the GNP. Certainly the chain of assumptions here is too long and tenuous to yield a very reliable estimate – the assumptions with respect to the velocity of circulation and the irrelevance of changing financial institutions (particularly credit cards) to people's preference to holding cash, are particularly insecure – but nevertheless subsequent professional comment on the article suggests that the estimate is quite credible.[6]

There may however be some rather less aggregated methods for estimating the underground economy. Let us return to the three categories of underground production identified in section 4.2. We have already referred to the possibility of tracing type A activities through national accounting surveys and through commercial estimates of occupational theft. Type B activities are very difficult to trace (and social science research in this area poses a particularly serious moral dilemma) – but this difficulty is of only limited significance since the final product emerges either in the formal economy or else as type C production. Type C activities are also difficult to estimate, but here we may be able to draw a suggestion from Gutmann's work. One identifying characteristic of the underground economy is certainly its reliance on cash. While we are unlikely to be able to persuade workers in the underground economy to identify themselves in response to survey questions, it should certainly be possible to ask whether people have paid for certain commodities *in cash*, whether they were asked to do so, and whether they were paid any discount for so doing, and use these answers as an indicator of underground production.

One pioneering study along these lines has been carried out in Detroit[7] by Ferman, Berndt, and Selo. (FBS). Their research had two phases. First, a group of anthropology graduate students spent some months collecting ethnographic data on the informal economy (with an emphasis on 'hidden production'), compiling through loosely structured interviews, notes on work histories and social circumstances of workers in the informal economy. Information from this source enabled the researchers to identify the sorts of tasks carried out by the 'hidden economy'; the second phase used these results in the context of a formal questionnaire – to be completed, not by the producers of hidden economy products but the consumers – asking about the source of supply of a very wide range of different personal and domestic services.

One of the most significant results stemming from the first phase concerned the identity of workers in the hidden economy. Contrary to the expectations of writers such as Gutmann, who sees the hidden economy as an *alternative* to formal employment, hidden work emerges from this study as typically a *complement* to formal work. Workers would frequently be found in 'irregular' employment because of low formal wages or because of the intrinsic satisfactions of the work, but not because of the lack of any sort of formal job; 'irregular' workers who were formally unemployed were in the minority.

But for our present purposes, the results of the second phase are more important. The researchers considered the sources of supply of a wide range of different personal and household services.[8] They found that only 30 per cent of these items of service were purchased from the formal economy – as against 60 per cent from household or communal sources, and 10 per cent from the 'irregular' economy. We must certainly use these statistics cautiously: the questionnaire was of a random sample of Detroit households, so the results need not necessarily throw much light on behaviour outside that city: and the study considers only the occurrence of a service, and not its value, so that in terms of value households may have been more dependent on the formal economy than these numbers would suggest. But, assuming away these difficulties for a moment, these numbers give us some feel for the scale of the informal economy as a whole: if, for the sake of argument, we assume that 10 per cent of GNP in the USA is spent on purchases of household services from the formal economy, then it would appear that the overall informal provision of services to persons and households, including hidden, household, and communal provisions, constitutes an addition of about 23 per cent (i.e. $\frac{7}{3} \times 10$ per cent) to GNP. And since the Michigan researchers leave out of account many household services that should be included within the terms of our present discussion (regular housework, cooking, cleaning, some car-driving), the actual addition to GNP from these services is probably considerably higher. However, only one seventh of this addition is due to the hidden economy – the rest coming from unpaid domestic production – so the data yields an estimate of type C output from the hidden economy of 3 per cent or 4 per cent of GNP in the USA.

These sorts of estimates were not, however, the primary purpose of the Michigan group; and there are some rather more sophisticated procedures for calculating the value of household services. There are in the literature two different ways of making this sort of imputation. (1) calculating the 'opportunity cost' – the cost of earnings forgone – to the *individual* engaged in informal production, and (2) calculating the cost of employing someone, either a specialist in the particular task, or a generalized housekeeper, to do the work. Hwrylyshyn, reviewing a number of sources (mostly using the second procedure) comes up with a range of estimates of value of household output ranging around 30 per cent to 35 per cent of measured GNP.[9]

Both of these procedures have some fundamental objections standing against them. Procedure (1) involves some rather doubtful assumptions about the relative informal productivities of occupants of different formal jobs. The underlying assumption of the 'opportunity cost' approach is that people choose the amount of time spent in different activities so that the marginal value of time spent in each activity is a constant; so the value of time spent in informal work must be equal to (or in the case of the formally unemployed, greater than) the individuals' marginal wage rate in the formal economy. The valuations therefore reflect (at best) personal evaluations of the value of the informal production rather than the interpersonal valuations normal in National Product calculations. The effect is to build in to the calculation an assumption that informal productivity is positively correlated with formal earning power – a most unlikely situation. This procedure alternatively makes a most doubtful macro-economic assumption – that the unpaid workers could be somehow absorbed into the formal economy without affecting the pattern of wage rates, again, most unlikely. So, whether it is assumed that the imputed opportunity cost reflects personal evaluations of informal production or actual wages forgone, this first procedure seems somewhat unreliable.

The second procedure also makes doubtful assumptions about the productivity of informal work. From personal experience, I would think it unlikely that an hour spent by a professional academic research worker in building bookshelves, is used as efficiently as an hour of the time of a professional carpenter at the same task.

It might also be objected that informal production involves the joint generation of what might be referred to as *direct* and *indirect* utilities. Making the bookshelves is itself a pleasurable activity to me (probably more so than is the same activity for the carpenter), and this direct utility (essentially consumption rather than production) should be accounted for.

But there is a more fundamental objection to this procedure. The chief reason that we are interested in the informal economy is its changing relationship with the formal. If the formal and informal economies were both to grow at the same rate, then for most purposes we could use the growth of the formal sector as a proxy for the growth of the economy as a whole. The second procedure is, however, inappropriate for making such comparisons between the two sectors over time, for one simple reason: service wages have, over the past decades, risen rather faster than average wages, so that if we value informal output by the wages of equivalent formal economy service workers, we *build into* our calculation a rising value of informal output which is purely an artefact of changing relative wages in the formal economy.

Following the line of argument in this book, it is possible to see a third approach to the problem: the construction of hypothetical household production functions from which household output might be synthesised. We have, both for the UK and for the USA, information on the change in the stock of household capital.

We have, for both countries, data on changing patterns of time use over time (for the USA this data was collected in comparable form in 1965 and 1975 by the Institute of Social Research at Michigan University; for the UK we have the comparative data for 1961 and 1974/5 constructed at SPRU, University of Sussex, from the diaries collected by the BBC). Using these data within a production function approach would yield some problems in estimating the *absolute* value of informal output – but it would give a better indication than either of the other procedures as to the *rate of growth* of informal output relative to formal. These three procedures cover mainly unpaid household and communal production, and will not contain estimates of hidden production. So to get a complete picture of the scale of informal production, it will be necessary to combine the time budget/capital endowment approach for domestic and communal production with the 'consumer survey' approach of Ferman, Berndt and Selo for hidden production.

4.5 The Distribution of Work Within and Between Households

So far in this chapter we have discussed the development of the informal production of services and not its *effects*. But of course the sorts of social innovations described here have important social consequences, particularly with respect to the division of work within and between households; in this final section we shall very briefly identify some of the possible negative effects of the growth of informal production.

We have identified a number of locations of productive activity outside the money economy – households, community institutions, the 'hidden economy'. In an ideal world, everyone, or almost everyone, would have responsibilities for a certain amount of each of these sorts of work. In such circumstances, most of the problems outlined in the following paragraphs need not trouble us. But unfortunately work is not distributed in this way; some people bear a very heavy and undesired responsibility for household production, a responsibility which may be added to by the processes of social innovation we have been discussing. Others (or sometimes the same people) have limited access to paid employment, a problem in itself which may also cause problems for the informal production activities of their households.

Household work strategies may be considered to have two components: decisions about the mode of provision of services (i.e. which services are to be bought, and which are to be produced within the household) which in turn determine the household's overall balance between paid and unpaid work; and decisions about which members are to be responsible for which sorts of tasks. Unpaid domestic work has been (and as we shall see in Chapter 9, still is) generally a segregated female task. Where paid employment has in general a complementary pattern of male segregation, female specialization in domestic work does not necessarily have any particular adverse consequences.

But the consequences of the sexual segregation of tasks do become serious at the point at which the segregation begins to break down. If a wife who has previously borne the full responsibility for the household work obtains a paid job, she does not, in general, reduce her unpaid work in proportion to her increase in paid work, nor does her husband significantly increase his contribution to domestic production. Under these circumstances, as time budget studies in a wide range of different cultures have shown, very considerable disparities between husbands' and wives' working times may develop. As a consequence of their 'dual burden' – paid and unpaid work – employed wives with full time jobs may perform less well in their jobs than their male colleagues. Or they may be forced to seek part-time jobs, which are in general worse paid, and offer less opportunity for career advancement. Rates of female participation in the paid work-force have been increasing very considerably; sexual inequality in work times may be developing into a major social problem throughout the developed world.

Now consider the implications of social innovations which increase the range of unpaid productive activities within the household. As the range of activities increases, we might assume, the amount of time spent in domestic production may also increase. If domestic (and communal) production remains a female segregated activity, then the disparity between men's and women's total work time may be further increased – the more so because of the progressive decline in the average weekly hours of full time employment. In fact, as we shall see in Chapters 8 and 9, the prospects are not really quite as dire as this argument might lead us to suspect – but nevertheless it is an issue that must be taken seriously.

We must also consider the problem of the distribution of work between households. In particular, unemployment means that some households have access *only* to informal work. The whole argument of the preceding chapters is that households increasingly acquire their services by a *combination* of unpaid work with capital goods and materials purchased with money earned in paid employment. Households without formal employment may therefore find that their abilities to engage in informal production activities are also impaired. And the more that a particular society provides its final services on an informal rather than a formal basis, the more expensive are its formal services. So the process of 'informalization' of production does not necessarily alleviate the burden of unemployment. And while unemployment *may* be partially compensated for by 'hidden' employment, or by work on the margins of the formal economy (Type C employment) – this is in general a very insecure means of livelihood.

So, while recognizing the importance of the growth of informal production, we should certainly not see it as any sort of social panacea. On the contrary, it poses a range of new problems – for sexual equality and for the alleviation of poverty. Informal production is an important part of any ideal world, of course, with the different sorts of work evenly distributed, within households and between them; we are still left with the question of how to *achieve* such an equitable distribution of work.

NOTES

[1] C. G. Homans, *Social Behaviour: Its Elementary Forms*, London, Routledge and Kegan Paul, 1962.

[2] S. Henry and G. Mars 'Crime at Work: The Social Construction of Amateur Property Theft', *Sociology*, Sept. 1978, No. 12.

[3] In fact part of the work done in categories B and C may be recorded; some workers may enter their category B and C earnings in their tax returns, and the income *may* appear in the spending recorded in household budget surveys. For a discussion of these shadows of underground economic activities in the official accounts, see Michael O'Higgins, 'Tax Evasion and the Self-Employed: an Examination of the Evidence', *British Tax Review*, No. 6, 1981.

[4] Theprimary reference here is to the theory of development expounded in K. Polanyi, *The Great Transformation* (Boston, Beacon Press, 1957); see particularly Chapters 4 to 10. For ease of exposition, I have combined Polanyi's two categories 'redistribution' and 'reciprocity' into 'generalized exchange'. This view of the social changes ('monetarization') accompanying economic growth remains an important theme in the conventional model of development; see, for example, Chapter 6 of F. Hirsch, *Social Limits to Growth*, London, Routledge and Kegan Paul, 1977.

[5] These are Peter M. Gutmann 'The Subterranean Economy', *Financial Analysis Journal*, Nov./ Dec. 1977, pp. 26–7, 33; and Edgar H. Feige, 'How Big is the Irregular Economy', *Challenge*, Nov./Dec. 1979, pp. 5–13. Gutmann's argument is described in the text of this chapter; Feige's method of estimation relies on a computation of the velocity of circulation of paper money calculated from the results of tests on the durability of dollar bills. Both articles are critically assessed in Vito Tanzi, 'The Underground Economy in the United States: Estimates and Implications', *Banco Nazionale del Laboro Quarterly Review*, No. 135, Dec. 1980, pp. 427–53.

[6] Estimates of the size of the underground economy in the UK include Adrian Smith, 'The Informal Economy', *Lloyds Bank Review*, July 1981; Kerrick Macafee 'A Glimpse of the Black Economy in the National Accounts', *Economic Trends*, Feb. 1980; A. Dilmot and C. N. Morris, 'What do we know about the Black Economy', *Fiscal Studies*, Mar. 1981; Christopher Johnson, 'Light on the Black Economy', *Lloyds Bank Economic Bulletin*, No. 38, Feb. 1982. The consensus seems to be that the overall size of the Black Economy in the UK is certainly no more than 5 per cent of GNP, and probably substantially less than this.

[7] *Analysis of the Irregular Economy: Cash flow in the Informal Sector.* A Report to the Bureau of Employment and Training, Michigan Dept. of Labour; L. A. Ferman, L. Berndt, E. Selo,; Institute of Labour and Industrial Relations; University of Michigan – Wayne State University.

[8] The full list of services considered was: exterior painting, wiring, tiling, carpentry, furnace repair, cement work, panelling, roofing, television repair, plumbing, interior painting, lawn care, hair care, babysitting, regular child care, help during illness, errands, automobile tuning, and loans.

[9] Oli Hawrylyshyn, *A Review of Recent Proposals for Modifying and Extending the Measure of GNP*, Statistics Canada, Dec. 1974; Oli Hawrylyshyn, *Estimating the Value of Household Work in Canada, 1971*, Statistics Canada, June 1978.

CHAPTER 5

Social Innovation, Diffusion and Waves of Development

5.1 Some Concepts Restated

The previous chapter develops a very general theoretical framework. From the next chapter onwards the concern will become very much more specific – to demonstrate, in an empirical fashion, that the concepts developed here are in fact useful, both in the description of past patterns of change in economic structure, and for thinking about future patterns. This chapter restates some of these concepts to serve as a basis for the empirical work. It will also take the opportunity to use these reformulated concepts as a platform on which to consider two new issues in a rather summary fashion: the 'diffusion of innovative products' and the connected notion of 'waves of innovation'.

It will be helpful at this point to introduce some formal definitions of the key concepts.

'Service functions' – these are the basic categories of need which are satisfied by economic activities (e.g. nutrition, housing, entertainment, medicine). We can specify a comprehensive list of service functions such that each element of final expenditure may be allocated to one or other of the elements in this list.

'The mode of provision of service functions' – each service function may be provided in a range of different ways. So private motoring and public transport represent alternative modes of provision for the transport function, TV and cinema alternative modes of provision of entertainment.

'Modal split' – is the statistic which indicates the distribution of provisions for a particular function between the alternative modes.

'Social innovation' – is used to describe the process whereby the modal split changes over time. (This is a special usage excluding some categories of social change that would be included in a broader application of the term).

5.2 Traditional vs Innovative Modes of Provision

In rather ideal terms, we might counterpose two alternative, competing modes of provision for some particular service function. The first is the traditional mode. It is labour intensive, it frequently involves the ultimate beneficiary in face-to-face contact with individuals paid to undertake activities on his or her behalf. The capital equipment involved is inconsiderable, and frequently very long-lasting. So the fixed costs of provision are low, while the marginal costs – which are chiefly labour costs – are high. What we have here, of course, is a traditional *final service industry* – a barber shop, or a restaurant, or a university.

Against this traditional mode of provision we may contrast the innovative mode. The cost structure of this second ideal type is rather different. The marginal costs are much lower relative to the fixed; capital costs are relatively high, while labour costs are low – in part because of the 'beneficiaries' contribution of their own unpaid 'informal' labour. Here we have *self-servicing* – DIY decorating, private motoring, or the Open University.

The reduction in paid labour inputs is achieved through three processes: *mechanization*, which reduces the physical work to be done; *automation* and improvements in materials technology, which simplify the tasks involved in the final provision of the service; and (this may alternatively be viewed as a more general statement of the second process) the *embodiment* of the knowledge and skills of service workers in *software*.

In the traditional mode of provision, the beneficiary simply buys, and directly consumes, a final service purchased from the money economy. By contrast, in the innovative mode of provision the beneficiary combines a range of factors in an activity that may be usefully viewed as a further process of production ('informal production'), taking place outside the conventional economy. We can identify five distinct classes of input, five 'factors of production' for this innovative mode of production:

– domestic machinery – 'consumer capital' – for informal production (e.g. cars, washing machines, TVs),
– infrastructure (e.g. electricity generation and distribution, road networks, broadcasting networks),

– materials, raw or semi-finished (e.g. petrol, electricity, washing powder).
– software and other intermediate services (e.g. equipment maintenance, TV programmes),
– informal labour, to combine and organize the other factors into the ultimate service functions.

These factors of production have varying degrees of importance in the various informal production processes. In the innovative mode of provision of domestic services, for example, software is very largely absent (or, alternatively, built into the machinery), and the low marginal costs are for the most part a consequence of the very considerable quantity of informal labour used. By contrast, in entertainment functions the innovative mode (TV, recorded music, etc.) involves

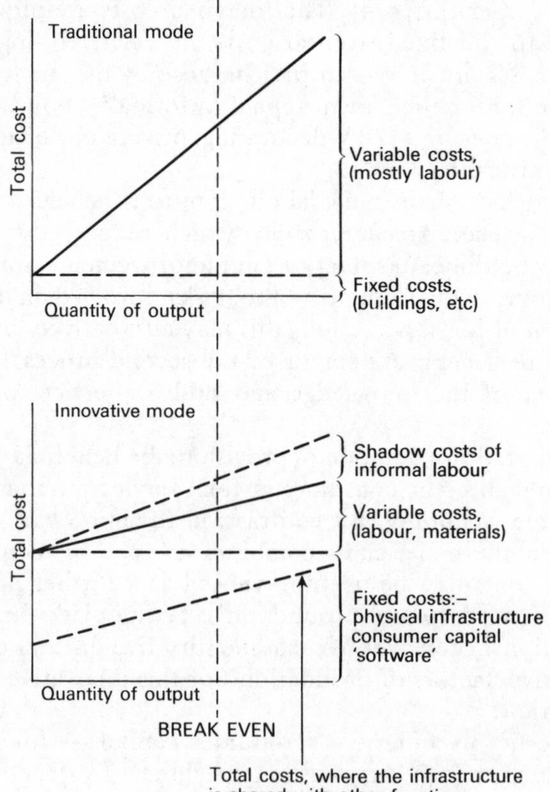

FIGURE 5.1 *Competing modes of provision*

virtually no informal labour but relies very heavily on software. But in general, at this rather abstract level of discussion, we can characterize the cost structures of the contrasting modes of provision as in Fig. 5.1.

Over time individuals and households may change their choice of mode for the provision of a particular service function. This change in the 'modal split' may be assumed to reflect the effective costs and quality of provision by the various alternative modes, which (following the arguments in Chapter 3), may in turn be viewed as being determined by:

- the extent of collective investment in infrastructure appropriate for innovation in particular service functions.
- the reduction in the cost of the consumer durables and materials used in the innovative modes of provision (i.e. the extent of process innovation in the manufacturing sector),
- technical improvements in the capacity and quality of domestic machinery (i.e. improvements in the efficiency of energy conversion and information processing).

Where such changes are taking place, and under the condition that the consequent improvement in the efficiency of informal production of services is not matched by an equivalent improvement in the performance of the final service industries, the modal split may be expected to shift towards the innovative modes of service provision. People satisfy their needs for particular service functions less through the direct purchase of final services, more by the purchase of goods, materials and intermediate services.

5.3 Modal Split and Product Diffusion

So we have a proposition about the effect of technical change on life styles that can be directly translated into an empirical, testable, form. If social innovation is taking place within a particular 'final service function' we should be able to observe a particular sort of *change in the pattern of final demand.* In the next chapter we shall consider actual data for several EEC states which does indeed demonstrate the expected proportional shifts of final expenditure from final services to other commodities. But before turning to the real statistical material it will be helpful to further develop our theoretical apparatus.

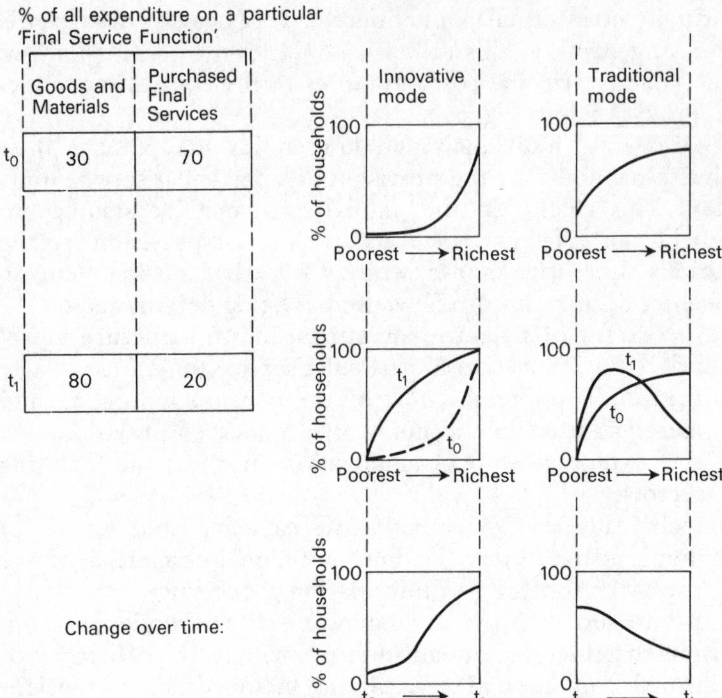

FIGURE 5.2 *Modal split and the diffusion of innovations*

Consider Figure 5.2: let us imagine that the modal split statistics refer to the final service function of transport, and that the first time period is 1930. Then we would have, in our invented statistics, 30 per cent of all expenditure devoted to motorcars and ancillary commodities, 70 per cent of expenditure devoted to the purchase of bus and train trips. Given this data we would expect to find distribution curves for the two modes such that most private motoring trips were taken by the very richest households, and that most of the public transport trips were made by the poorer households.

As time passes we invest in infrastructure (roads), and technical change both reduces the cost of motor cars (by improvements in production technologies) and increases their reliability. Over the same period, because of lack of productivity growth, public transport costs rise relative to other commodities. As a consequence, by the second time period (perhaps 1950) the modal split has reversed itself – 20 per cent of all transport expenditure now being devoted to public transport and 80 per cent to private motoring. And again we may guess that this reflects a change in the distribution of trip-modes across households, with private motoring becoming much more evenly distributed, while public transport becomes disproportionately the preferred mode of the poorest households.

So we can use our techno-economic model of social innovation to explain a visible and familiar change in life-style. The argument throws light on more than just social change, however. It also goes some way towards explaining a familiar economic issue, the diffusion of new products.

The S-shaped curve, with an initial slow growth of the market for an innovative product, which after a time achieves 'take-off' into a period of high growth of demand, which itself is succeeded by a gradual slowdown as the market saturates, is a commonplace of technical economics. But what allows the product to 'take-off'? What conditions the ultimate saturation level? In our model of social innovation we have a quite general answer to these questions. A product will achieve take-off when, in combination with other products, and with particular infrastructural provisions, it can be used to produce a particular final service function more effectively than alternative systems of production (i.e. modes of provision) do. The product reaches market saturation when all the households that possess the facilities for or inclination towards employing the innovative mode already do employ it, while those who still use the traditional mode will always continue to do so (either because of personal disabilities, or dissatisfaction with the quality of service from the innovative mode). This line of argument clearly has rather general implications for the relationship between technical innovation and economic development.

5.4 Economic Growth and Social Innovation

We can move from micro-economic considerations to more macro-economic ones. The model provides us with a plausible basis for understanding the reasons for long-term irregularities in rates of economic growth – for explaining why some decades, or even longer periods, might have better *potential* for economic growth than others. To put the argument at its simplest, it may be that at some historical junctures there exists a bundle of interrelated technologies which, because of lack of appropriate infrastructure and suitable social forms, are not and cannot be translated into social innovations, and so economies are stagnant. At other junctures new infrastructures are being constructed, social forms are sufficiently malleable to enable innovation, and economies are buoyant.

Let us consider how this emerges from the social innovation model. The model suggests that the development and diffusion of new products depends on two sets of circumstances: a 'bundle' of new technologies and a new sort of infrastructural provision.

The bundle of new technologies, on past evidence, involves technical advance both in the design of new products and in production processes. On one hand it will tend to include a set or system of mutually potentiating inventions which together provide new capacities for mechanization, automation, or other technical change in the mode of provision of services. On the other, it will include improvements in the efficiency of production of machinery, particularly domestic machinery, which makes consumer capital goods cheaper and more accessible to households.

New infrastructural provisions have a number of characteristics. They are very expensive, and are therefore only likely to be developed if they provide for a range of different services. (So, for example, the electricity supply system provides for a broad range of different domestic and entertainment services.) But though they are general-purpose, their range of application is nevertheless finite – particular sorts of infrastructure will be inappropriate for innovation in particular sorts of services. (The electricity supply system for example is not a particularly suitable medium for person-to-person telecommunications.) And by nature, infrastructural provisions are, to use the economists' jargon, 'lumpy' – to be effective, they must be diffused and generally available across the society. As a consequence of this, investment in one particular sort of expensive infrastructure may have the (at least temporary) effect of 'crowding out' investment in other sorts of infrastructure.

With these considerations in mind, we can propose an extension to the conventional Schumpeterian model of the relationship between technical change and economic growth. Schumpeter's model,[1] reduced to its broadest essentials, proceeds as follows: an invention (or set of inventions) gives rise to a radically new and clearly desirable product, which develops a rapidly growing market, and provides high rates of profit for its entrepreneurs. This radical innovation attracts more producers, who allow a very considerable growth in the volume of output of the new product, but inevitably depress profits somewhat. At the same time, the initial radical innovation inspires other, less radical innovations, which create further new markets, perhaps fuelled (through the 'multiplier effect') by the growth of the market for the initial radical innovation. So there is a wave of growth, stimulated by an initial important new product, followed by other lesser waves; after a time markets become saturated, fewer and less promising new products are marketed, and the rate of profit falls, so setting the stage for the next radical innovation.

This model is not in itself, however, able to account for long term historical differences in the growth potential of economics. Why should growth be so much easier in the 1950s than in the 1970s? Why should radical innovations not be evenly spaced in history, perhaps giving short-term fluctuations, but evening out when viewed from decade to decade?[2]

The social innovation model does suggest some answers to these questions. Once we see the diffusion of particular new products as *dependent on their combination with other new products and with appropriate infrastructure* so as to provide a viable alternative mode of provision for some service function, then we have a mechanism which may encourage the diffusion of a bunched group of new products at one historical juncture and inhibit or prevent the diffusion of new products at another. We can extend the Schumpeter view as follows:

1. We might start in a stagnant economic climate where there is no substantial innovation, but nevertheless thwarted innovative *potential*. There is a developing cluster of inventtions which do not get embodied in new products because there is no appropriate infrastructure – and, perhaps even more important, no widely shared vision of how these inventions might be combined and used in new modes of provision of services.
2. The cluster of innovations grows, and it becomes clear that

particular sorts of social innovations might be credible and feasible given appropriate infrastructure. At this stage, particular demonstration projects might publicize the potential of the new wave of innovations.

3. As the economic crisis deepens, there is growing pressure for new investment to generate jobs; since it at once provides scope for social innovation (and new products) and affects the demand for the new products.
4. Once the new infrastructure is substantially present, a wave of new products develops to exploit it. And as, after a period, the finite capacities of the particular class of infrastructure become fully utilized, the economy stagnates – but by this time a new bunch of thwarted innovations may have developed. So the cycle is repeated.

We shall argue that we can see in the Europe of the 1950s and 1960s, the middle stages of one such cycle. The new infrastructural facilities were the electricity generating networks, the burgeoning road systems, the broadcasting networks. The new technologies were the fractional horse power electric motor, valves and then solid-state semi-conductors, and the internal combustion engine. These in combination allowed social innovations in domestic service, transport, and entertainment functions which transformed both life-styles and economic structures.

We shall argue that as we move into the 1980s, a period of economic stagnation, we can now see the prospects for the next wave of innovations. Its necessary infrastructure is a broad band-width telecommunications network. Its critical technologies are clustered around the microchip computer and cheap data storage media. And it will consist of a wide range of new modes of provision of service functions – in entertainment, information and marketing, education, and perhaps a little further away, personal and medical care and counselling.

NOTES

1 In J. A. Schumpeter, *Business Cycles* (2 Vols.), McGraw Hill, 1939.

2 There are two current schools of thought concerning the relationship

between 'bunches' of major innovations and long term differences in levels of economic activity. Gerhard Mensch, in *Stalemate in Technology*, Ballinger, 1979, argues that depressions stimulate fundamental technical innovations. Closer to the position outlined in this chapter, however, is the argument in C. Freeman, J. Clark and L. Soete, *Employment and Technical Innovation*, Frances Pinter, 1982, that the diffusion of clusters of related innovations is an important mechanism in the *recovery* from major depressions.

CHAPTER 6

The National Accounting Model 1: From Consumption Patterns to the Industrial Distribution of Employment

6.1 Introduction

The central theme of this book is the interrelationship between social innovation and economic structure. The conventional wisdom is that the 'service' sectors of the economy will grow – but what *are* services? We can identify at least five distinct meanings, applying respectively to particular categories of occupations, industrial employment, industrial output, consumption commodities, and what we referred to in previous chapters as 'final service functions'. The relationship between these various sorts of services is a complex one. Consuming more of a particular final service function does not necessarily lead to more demand for a service commodity – we can spend more on clothes cleaning, but buy more washing machines and less laundry services. But increasing demand for manufactured commodities may nevertheless enlarge the output of service industries, by increasing the demand for the products of specialist design, marketing, advertising, or factory-cleaning firms. Even if the demand for service industry output falls as a proportion of GNP in real terms, low labour productivity might mean that the sector's proportion of total employment might nevertheless grow. And finally, manufacturing industry itself provides many 'service' jobs – in management, technical clerical, catering, selling, cleaning, and other occupations.

In the present and the following chapters, we shall consider the complex set of relationships between these various different sorts of economic and sociological categories. We shall develop a systematic framework for relating 'life-style' and economic structural data. The argument is unavoidably somewhat complicated, so by way of introduction, this section will present a very much reduced version of the argument; those readers who are antipathetic to discussions of statistical material will find the gist of the analytical points here – and a brief summary of the conclusions of the empirical discussions in Sections 6.2 to 7.4 will be found in Section 10.3.

Let us return to our hypothetical transport example. Of 1,000 units of expenditure on the transport 'service function' in 1930, let us say that 300 units go on cars and their associated infrastructure, and 700 units are spent on purchasing transport services like train or bus trips. So, we would say, 70 per cent of transport related expenditure purchases final service commodities, and 30 per cent goes on goods. What does this mean for the output of the service industries relative to the other sectors?

It certainly does *not* mean that service industries produced seven-tenths of all transport-related output. Consider: a large part of a service industry's output relies on manufacturing production. Trains must be built somewhere, coal for steam-raising must be mined, steel rails must be forged; part of the final cost of a train ticket must be accounted for by the 'value added' by primary and manufacturing industries. Similarly – though perhaps to a lesser extent in 1930 than at present – the 30 per cent of transport-related final expenditure spent on goods does not go entirely to primary and manufacturing industries. The car manufacturers presumably employed advertising agencies to promote their products, and appointed dealers to sell them. If, to continue our hypothetical 1930s example, 40 per cent of all consumers' expenditure on final services went to pay for value added by primary or manufacturing industries, and 8 per cent of final expenditure on goods went to pay for the production of service industries, then (as the first part of Table 6.1 makes clear) the 30/70 proportional split between final expenditure on goods and services translates itself into a 55.5/44.5 split between primary and manufacturing industries on one side and service industries on the other. Service commodities may be the predominant proportion of final expenditure, yet still make up a minority of industrial value added.

Assume now that employees in both sectors add on average one unit of value per year; the thousand units of expenditure on transport will then provide for 1,000 transport-related jobs, distributed as 555 jobs in primary and manufacturing industry, and 445 in service industries. This is not quite the end of the story, however. Manufacturing industries employ

TABLE 6.1 *Hypothetical relationship between final demand, value added and employment, transport, 1930 and 1980*

1. 1930

INDUSTRIAL CATEGORIES	COMMODITY CATEGORIES: Goods (1980 Prices)	COMMODITY CATEGORIES: Services	Industrial Value Added	Value Added Per Head	Industrial Employment (number of employees)	OCCUPATIONAL CATEGORIES: White Collar	OCCUPATIONAL CATEGORIES: Manual
Primary, Manufacturing	275	280	555	1.00	555	56	499
Services	25	420	445	1.00	445	120	325
All Transport Expenditure	300	700	1000	All Transport Employment	1000	176	824

2. 1980

INDUSTRIAL CATEGORIES	COMMODITY CATEGORIES: Goods (1980 Prices)	COMMODITY CATEGORIES: Services	Industrial Value Added	Value Added Per Head	Industrial Employment (number of employees)	OCCUPATIONAL CATEGORIES: White Collar	OCCUPATIONAL CATEGORIES: Manual
Primary, Manufacturing	2440	320	2760	3.00	920	276	644
Services	760	480	1240	1.82	680	262	418
All Transport Expenditure	3200	800	4000	All Transport Employment	1600	538	1062

TABLE 6.1 *continued*

3. Proportions in Each Category

	FINAL COMMODITIES		INDUSTRIAL VALUE ADDED		EMPLOYMENT BY INDUSTRY		EMPLOYMENT BY OCCUPATION	
	1930	1980	1930	1980	1930	1980	1930	1980
	(% of final expenditure)		(% of value added)		(% of industrial employment)		(% of occupational employment)	
Goods (Manual Occupations)	30	80	55.5	69.0	55.5	57.5	82.4	66.4
Services (White C Occupations)	70	20	45.5	31.0	45.5	42.5	17.6	33.6

some white-collar 'service' workers (say 10 per cent of their total employment for the sake of our example). And service industries employ some manual workers – and in the particular case of transport, a *majority* of manual workers (say 73 per cent). On these assumptions, the 55.5/44.5 split between non-service and service industry employees becomes an 82.4/17.6 split between manual and white-collar occupations. In short; the 70 per cent of transport-related final expenditure that goes on service commodities becomes 45.5 per cent of output and employment in service industries, which leaves in turn 17.6 per cent of transport-related employment in service occupations. In other cases of course the relationship might go in the other direction; the 'communications' function for example, might have a majority of consumer expenditure going on goods, with the majority of the associated employment being in white-collar occupations. The point is that, just as there is no simple relationship between the kinds of functions (e.g. transport, domestic services, entertainment) being met and the kinds of commodities (goods or services) purchased to meet them (because of alternative possible modes of provision), so there is no simple relationship between the kinds of commodities purchased by consumers and the sorts of jobs these purchases support.

And just as these relationships are not *simple*, they are also not *necessary*. They may change over time. In the previous chapters we have already come across two different categories of change. As societies get richer, their members may be expected to shift their expenditure proportionately towards the satisfaction of more 'luxury' functions (the Engel's Law effect). Thus, taking transport to be such a 'luxury', we might posit a fourfold increase (in real terms) of transport expenditure in our hypothetical example, from 1,000 units to 4,000 units in 1980. As time passes, the modal split may also change (the social innovation effect). Let us say that it changes from a 70/30 split in favour of services, to 20 per cent of expenditure on services and 80 per cent on goods. These two changes mean that in 1980 expenditure on transport-related services is pretty much unchanged (800 units, as against 700 in 1930) where transport-related goods have increased more than tenfold (300 units to 3,200).

Now, if all the other relationships in our model remain unchanged, the new pattern of final expenditure would lead to a decline in the service sector's proportion of transport-related value added and employment from around 45 per cent to 25 per cent, and reduce transport-related white collar employment from around 18 per cent to 14 per cent of the total. But, over this long period, we would expect a number of the other relationships to change.

We would expect that, as a result of the division of labour between industrial sectors, the proportion of 'intermediate' services (like technical consultancy or factory cleaning) contributing to the final production of goods should increase. In the hypothetical case described in Table 6.1 this proportion grows threefold, partially, but not entirely, compensating for the decline in final demand for services. Instead of 14 per cent of transport-related value addedcoming from the services, then, we have about 31 per cent in 1980 – still a decline from the 1930 proportion, but only half as large a decline as would have occured if demand for 'intermediate' services had not grown.

We would expect changes in the relative productivity of the industrial sectors. If this remained the same as in 1930, then the 31 per cent of industrial value added coming from the service sector would support 31 per cent of all transport-related jobs. But as we have already observed, productivity in the service sector tends to grow more slowly than elsewhere in the economy. In our example, we assume that labour productivity rose threefold in manufacturing industry, but only by a factor of 1.82 in the services. The consequence of this assumption is that employment in the services, in our example, falls much more slowly than the service sector's proportion of value added; in fact the employment proportion is only reduced by three points, from 45.5 per cent to 42.5 per cent of all transport-related jobs.

Finally, as a result of the increasing technical complexity of production processes, and the consequent division of labour *within* industries, we would expect a growth in the proportion of workers in each sector employed in white collar or other non-manual occupations the consequence of this for our example is that overall the proportion of white-collar workers approximately doubles between 1930 and 1980, from 17.6 per cent of all transport-related employment, to 33.6 per cent.

So we have an extremely complicated set of relationships.

Engel's Law, in a period of economic growth, means that expenditure on the 'transport' function grows very considerably. But the social innovation effect, which changes the 'normal' mode of transport from collectively provided services to the private car, means that the proportion of final expenditure devoted to transport services declines very substantially. The growth in intermediate transport-related demand for services partially compensates for this, as does the relatively low productivity growth in the service sectors, so that the proportion of transport-related employment in service industries hardly declines. And within-industry occupational specialization, under these circumstances, means that the proportion of transport-related employment in service occupations virtually doubles. Declining final demand for services is coupled with substantially increasing demand for service workers.

6.2 The Accounting Framework

The discussion in the previous section gets us most of the way to an accounting framework which exhibits the relationship between social innovation and the division of paid labour. The one missing element is directly related to social innovation itself. We must find a means of generalizing the argument; so far we have considered the consequences of a change in the mode of provision of just one 'service function' (transport). We must now widen the analysis to include *all* the categories of final expenditure in the economy – and also all those final commodities (like medicine and education in the UK) produced by collective bodies and distributed without direct change.

We are accustomed to thinking of the final products of the economy in the form of a one-dimensional list; the economy produces so much food, so many cars and washing-machines, so much transport and laundry services, and so on. The total product of the economy is the total of each of these items. For our purposes we need to move to a slightly more complex presentation; instead of a one-dimensional list of categories, we must describe each item of final expenditure by two different sorts of characteristics. First, as we did in our transport example in the previous section, we must classify each item by the sort of commodity that it is. In the simplest case, we need only say whether it is a 'good' or a 'service' – though, as the argument develops we will need to construct a more sophisticated set of categories. Second, we need to identify the *sort* of need – the 'final service function' – that the particular commodity is satisfying. So, just as we had goods and services satisfying transport functions, we can have goods and services going ultimately to satisfying entertainment, or education or medical functions. We have to develop a system of categorization of final expenditure such that each item can be classified according to *the sort of commodity it is*, and *the sort of final need it satisfies*.

This final expenditure accountancy can actually be implemented within some national accounting systems (the UK system is not particularly suitable, but the UN and EEC systems, which embody the sort of functional classification suggested here, are appropriate for this purpose). Table 6.2 gives an example of such a system of accounting definitions, developed from the European System of National Accounts. It identifies ten separate categories of 'need' or 'final service function' (food, shelter, domestic entertainment, transport, educational, medical, other government, defence, and others); the list is, of course, arbitrary, the classification might be made either more or less aggregated according to the particular purpose it is to be put to. And it identifies three different sorts of commodity: primary and manufactured (in which are included intermediate services purchased by households), marketed services (final services directly purchased by households), and non-marketed services (those final services provided to households, in the main part without any direct payment, by government or charitable agencies); again, this categorization, though appropriate for the present purposes, might be very considerably disaggregated. Here, to continue with our example, those expenditures classified by the European System of Accounts as 'personal transport equipment' and 'operation' fall into the Transport/Primary and Manufactured Goods category, purchased transport services into the Transport/Marketed Services category, and public expenditure on roads etc. go into Transport/Non-Marketed Services.

Another substantial advantage to the European System of Accounts is that it supplies industrial employment statistics

TABLE 6.2 *Purposes of household and government final expenditure, classified by function and by the type of commodity*

'Function' Classification	'Commodity' Classification Primary and Manufactured Goods	Marketed Services	Non-Marketed Services	References in parentheses to classifications of the European System of Integrated Economic Accounts (ESA)
A. Food, Drink, Tobacco	Food, Drink, Tobacco (D1)	–	–	
B. Shelter, Clothing	Rent, Fuel and Power, Clothing and Footwear (D2, D3)	Personal Care and Effects (D81)	Housing and Community Amenities (Sewers etc) (G6)	
C. Domestic Functions	Furniture, Furnishings, Appliances, Utensils and Repairs to These (D41 to D44)	Household Operation and Domestic Services (D45, D46)	Social Security and Welfare Services (G5)	
D. Entertainment	Equipment, Accessories, and Repairs to These, Books etc (D71, D73)	Entertainment, Recreation, Cultural, Hotels, Cafes, etc Packaged Tours (D72, D83, D84)	Recreational Culture and Religious Services (G7)	↑
E. Transport, Communications	Personal Transport Equipment and Operation (D61, D62)	Purchased Transport and Communications Services (D63, D64)	Roads, Waterways, Communications, and their Administration Subsidies (G8.5, G8.6, G8.7)	Functions Provided Mainly by Households
F. Education	–	Purchased Education (D74)	Public Education (G3)	Functions Provided Mainly by Governments
G. Medical Functions	Medical and Pharmaceutical Products and Appliances (D51, D52)	Purchased Medical Services, Medical Insurance Service Charges (D53, D54, D55)	Public Health Services (G4)	↓
H. Other Government Functions	–	–	General Public Services, and Economic Services excluding Transport and Communications (G1, G8.1 to G8.4, G8.8)	
I. Defence	–	–	Defence (G2)	
J. Functions NES	Goods NES (D82)	Services NES (D85, D86)	Other Public Services NES (G9)	
	ESA Classifications and Coding of the Purposes of Final Consumption of Households. ESA 1979 Table 7. (The same as Table 6.1, SNA, UN, New York 1968).		ESA Classification and Coding of the Purposes of General Government. ESA 1979, Table 8. (The same as Table 5.3, SNA, New York, 1968).	

on a basis which is fully compatible with the expenditure and output figures. So it enables us to make estimates of the sorts of relationships discussed in the previous section for a number of different European countries.

Before we consider these relationships on an international comparative basis, however, let us look first at just one country. Table 6.3 presents in its simplest form the complete set of consumption, output and employment data for Belgium.[1] By working through this set of numbers we can give an overview of the working of the full accounting model – and see the relationship between social innovation and the distribution of employment. The 'function vector' gives the percentage allocation of household plus government spending to each of the final service functions.

In our model, we see final consumption as being determined by two different sorts of allocation decisions:

- decisions about the allocation of resources to particular final service functions;
- within each function, allocation of resources to particular sorts of commodities (on the basis of the desired 'mode of provision' of the particular service function).

As we argued in the previous chapters, over time, changes in relative prices, efficiency of domestic machinery, and levels of infrastructural provision, lead to changes in households' desired modes of provision (i.e. 'lifestyle' changes) which are reflected in changing patterns of demand for particular commodities.

In Table 6.3 'function vector' shows the change in Belgian expenditure on the various final service functions; as we would expect, the major differences between the two years are a reduction in expenditure on the more basic categories (i.e. 3.9 per cent less going on food) and an increase on the more sophisticated (3.5 per cent more on transport, education, and medicine). The matrix to the left of this vector gives the change in the proportion (again percentage) of the expenditure (in constant prices) on each function that is accounted for by the three commodity categories 'primary and manufactured goods', 'marketed services', and 'non-marketed services'. In the three relevant categories of final service function 'domestic', 'entertainment', and 'transport' we see the change in consumption pattern characteristic of social innovation. In

TABLE 6.3a *Belgium 1970*

Product Categories

Industrial categories	Primary	Manufactures	Marketed Services	Non-Marketed Services	Distribution of Value Added	Productivity Relative to Average	Distribution of Employment
1. Farming					3.6	0.73	4.9
2. Energy					5.2	2.50	2.1
3. Manufacturing					30.2	0.99	30.4
4. Construction					7.7	0.90	8.5
5. Marketed Services					41.5	1.13	36.8
6. Non-Marketed Services					11.7	0.68	17.3
					'Value Added Vector'	'Productivity Vector'	'Employment Vector'
Distribution of all final Expenditure by commodity	35.6	20.1	21.6	22.7	'Commodity Vector'		

Functional categories	Goods	Marketed Services	Non-Marketed Services	Distribution of all Final Expenditure by Function
A. Food	100	0	0	23.7
B. Shelter	92	8	0	13.7
C. Domestic	56	29	16	11.9
D. Entertainment	27	71	2	11.7
E. Transport	68	20	12	10.8
F. Eduction	0	0	100	8.6
G. Medicine	26	69	5	6.4
H. Other Government	0	0	100	4.3
I. Defence	0	0	100	5.6
J. Other	←8	5→	15	3.3
				'Function Vector'

TABLE 6.3b *Belgium 1976*

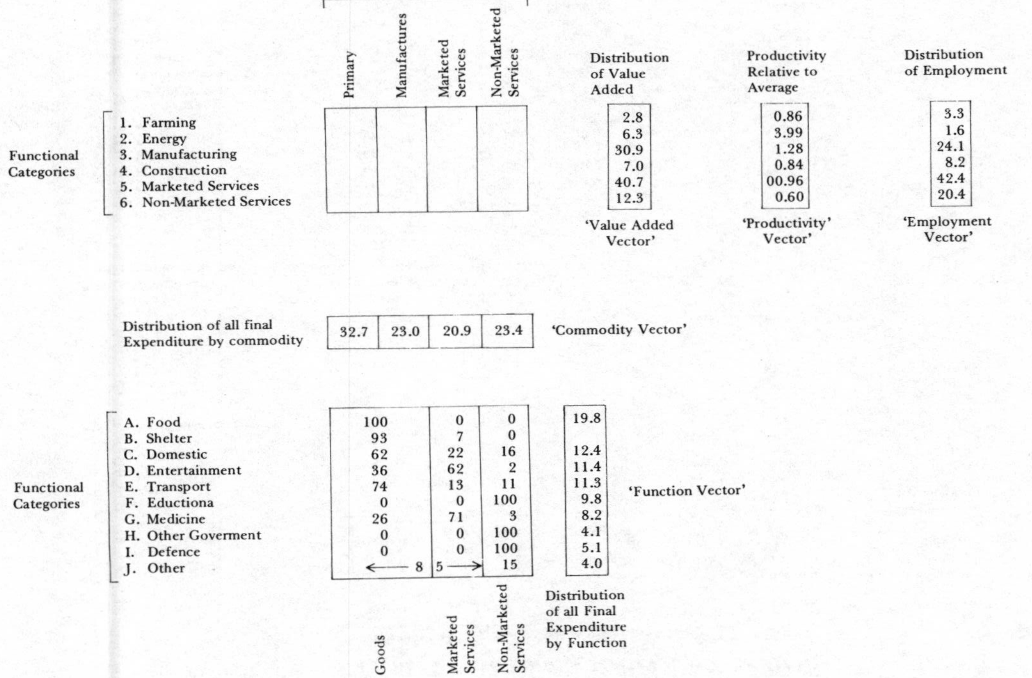

Product Categories

Functional Categories	Primary	Manufactures	Marketed Services	Non-Marketed Services	Distribution of Value Added	Productivity Relative to Average	Distribution of Employment
1. Farming					2.8	0.86	3.3
2. Energy					6.3	3.99	1.6
3. Manufacturing					30.9	1.28	24.1
4. Construction					7.0	0.84	8.2
5. Marketed Services					40.7	00.96	42.4
6. Non-Marketed Services					12.3	0.60	20.4
					'Value Added Vector'	'Productivity' Vector'	'Employment Vector'

	Primary	Manufactures	Marketed Services	Non-Marketed Services	
Distribution of all final Expenditure by commodity	32.7	23.0	20.9	23.4	'Commodity Vector'

Functional Categories	Goods	Marketed Services	Non-Marketed Services	Distribution of all Final Expenditure by Function
A. Food	100	0	0	19.8
B. Shelter	93	7	0	
C. Domestic	62	22	16	12.4
D. Entertainment	36	62	2	11.4
E. Transport	74	13	11	11.3
F. Eductiona	0	0	100	9.8
G. Medicine	26	71	3	8.2
H. Other Goverment	0	0	100	4.1
I. Defence	0	0	100	5.1
J. Other	← 8	5 →	15	4.0

'Function Vector'

TABLE 6.3c *Change: Belgium 1970–6*

Product Categories

Functional Categories	Primary	Manufactures	Marketed Services	Non-Marketed Services	Change in Distribution of Value Added ('Value Added Vector')	Productivity Relative to Average ('Productivity Vector')	Change in Distribution of Employment ('Employment Vector')
1. Farming					−0.8	1.18	−1.6
2. Energy					+1.1	1.59	−0.5
3. Manufacturing					+0.7	1.30	−6.3
4. Construction					−0.7	0.93	−0.3
5. Marketed Services					−0.8	0.86	+5.8
6. Non-Marketed Services					+0.5	0.88	+3.1

	Primary	Manufactures	Marketed Services	Non-Marketed Services	
Change in Distribution of all final Expenditure by commodity	−2.9	+2.9	−0.7	+0.7	'Commodity Vector'

Functional Categories	Goods	Marketed Services	Non-Marketed Services	Change in Distribution of all Final Expenditure by Function ('Function Vector')
A. Food	0	0	0	−3.9
B. Shelter	+1	−1	0	+0.2
C. Domestic	+6	−6	0	+0.5
D. Entertainment	+9	−9	0	−0.3
E. Transport	+6	−5	−1	+0.5
F. Eductiona	0	0	0	+1.2
G. Medicine	0	+2	−2	+1.8
H. Other Goverment	0	0	0	−0.2
I. Defence	0	0	0	−0.5
J. Other	0		0	+0.7

1970, for example, 56 per cent of expenditure on domestic function went on goods and materials, whereas in 1976 62 per cent was spent in this way; the proportion devoted to final marketed services declined by an equivalent amount. Similarly, the 'entertainment' function showed an increase in goods expenditure of 9 per cent of the total over the period, and a 9 per cent decline in the final service proportion, and the transport function showed a similar pattern.

This 'social innovation' (which as we shall see follows a quite regular pattern through the EEC) has the consequence that over the period we are considering, the demand for final marketed services actually *decreased* a little, while, as we see from the 'commodity vector' in Figure 6.2c – the demand for manufactured products increased by nearly 3 per cent as a proportion of all final expenditure. In Belgium this declining demand for final marketed services led to a similar small decline in the proportion of value added accounted for by the marketed services industry. This connection is not in fact a necessary one, however, since this industry also provides 'intermediate services' chiefly to manufacturing industry: as we shall see, in most European countries, growth in this intermediate demand counteracts the decline in final demand for services, so as to give a small increase in this industrial category's proportion of total value added in the economy.

And from the distribution of value added by industry, we can move, via the 'productivity vector' to the distribution of employment. Our Belgian example shows quite clearly that the growth in the proportion of employment accounted for by the two services is explained much better by the relatively low productivity growth in the services than by a proportional increase in demand for services – again, this is true, as we shall see, right across the EEC.

6.3 Europe in the 1970s

We can now turn to a multinational comparative analysis. Unfortunately the European System of Accounts has only published statistics for the 1970s – but this data is sufficient to allow us to demonstrate the consequences of our two effects.

Let us consider first the more conventional Engel's Law effect, by summing the categories in each row of Table 6.4 to

TABLE 6.4a *Final consumption by function*

A. *Proportions*

	Germany		Holland		UK		Belgium		Italy	
	Total		Total		Total		Total		Total	
	Early 1970s	Late 1970s	Early 1970s	Late 1970s	Early 1970s	Late 1970s	Early 1970s	Late 1970s	Early 1970s	Late 1970s
A. Food	21.4	19.7	20.0	19.2	21.8	19.4	23.7	19.8	33.4	30.5
B. Shelter	14.0	13.0	13.6	11.6	13.0	13.1	13.7	13.9	14.2	14.1
C. Domestic	10.8	10.6	10.6	9.0	7.6	8.2	11.9	12.4	7.2	7.4
D. Entertainment	11.7	11.9	8.2	10.0	16.1	16.9	11.7	11.4	12.3	13.5
E. Transport	11.4	13.2	7.5	10.2	12.1	12.7	10.8	11.3	9.8	10.2
F. Education	5.2	5.8	9.6	9.9	7.8	7.9	8.6	9.8	6.6	7.4
G. Medicine	9.8	12.1	8.8	8.9	6.4	6.8	6.4	8.2	4.6	5.6
H. Other Govt.	7.6	6.0	—	—	4.6	4.7	4.3	4.1	6.7	6.4
I. Defence	5.2	4.5	5.6	4.2	7.5	6.7	5.6	5.1	3.2	2.9
J. Other	2.9	3.2	16.1	17.0	3.1	3.7	3.3	4.0	2.0	2.0
Total	100.0	100.0	100.0	100.0	100.0	100.0	100.0	100.0	100.0	100.0

TABLE 6.4b

	G	H	IT	UK	B	ROW COUNTS —	0	+
A. Food	−1.7	−0.8	−2.9	−3.9	−3.9	5	0	0
B. Shelter	−1.0	−2.0	−0.1	−0.1	+0.2	4	0	1
C. Domestic	−0.2	−1.6	+0.2	+0.6	+0.5	2	0	3
D. Entertainment	+0.2	+1.8	+1.2	+0.8	−0.3	1	0	4
E. Transport	+1.8	+2.7	+0.4	+0.6	+0.5	0	0	5
F. Education	+0.6	+0.3	+0.8	+0.1	+1.2	0	0	5
G. Medicine	+2.3	+0.1	+1.0	+0.4	+1.8	0	0	5
H. Other Govt.	+1.6	(—)	−0.3	+0.1	−0.2	3	1	1
I. Defence	−0.2	−1.4	−0.3	−0.8	−0.5	5	0	0
J. Other	+0.3	+0.9	0	+0.6	+0.7	0	1	4

give us the total expenditure on each function. Table 6.4 shows the change in the distributions of all final expenditure for five European countries during the 1970s.[2] A quite clear pattern of change emerges. In spite of the relatively slow economic growth of the 1970s, expenditures in all of the five countries for which data are available show a decline in proportional expenditure on shelter; by contrast, four out of five show an increase in proportional expenditure on entertainment, and all increase their expenditure on transport. The more 'basic' categories are substituted for by the more 'luxury'. Similarly, the traditional 'basic' public services of defence and public administration proportionately decline, while the new luxuries of education and medicine grow. This is exactly the pattern of development we would expect from Engel's Law.

But this does not tell us the whole story; we must also look at changes in patterns of expenditure within each final service function. Table 6.5[3] shows the change in the split between household expenditure on goods and on services during the 1970s. No very clear pattern of change emerges for the more basic food and shelter categories (largely as a matter of definition – the very great majority of all expenditure on these categories is necessarily on goods and materials). But in the next three categories, domestic, entertainment, and transport functions, the split preponderantly shifts away from the purchase of final services, and towards the purchase of goods. This shift away from the purchase of final services is an indicator of the social innovation effect; it is precisely people buying less finished services from the formal economy, and instead buying goods as inputs to the informal provision of services. We might note that this data covers the 1970s, by which time markets for most of the basic goods meeting these three functions – cars, washing-machines, televisions etc. – were approaching saturation in Europe. It seems reasonable to presume that data for the 1950s and 1960s would show much larger changes in the same direction. (The remaining functional categories show no regular patterns; we shall return, in Chapter 10, to suggest why social innovation may be concentrated in particular functions over particular historical periods – for the moment, it will suffice to say that the available technologies and social and material infrastructure

TABLE 6.5a *Modal split, household provisions of service functions, Europe 1970 and 1979*

	(% of Expenditure by Function)		Belgium		Denmark		France		Germany		Holland		Ireland		Italy		UK	
			Goods	Services	Goods	Services	Goods	Services	Goods	Services	Goods	Services	Goods	Services	Goods	Services	Goods	Services
Basic Needs	A. Food	1970	100	0	100	0	100	0			100	0	100	0	100	0	100	0
		1979	100	0	100	0	100	0			1,000	0	100	0	100	0	100	0
	B. Shelter	1970	92	8	94	6	95	5			95	8	97	3	89	11	95	5
		1979	93	7	96	4	94	6			95	5	94	6	89	11	95	5
	C. Domestic	1970	67	33	67	33	68	32			82	18	70	30	61	39	74	26
		1979	74	26	75	25	75	25			82	18	74	26	62	38	74	26
	D. Entertain.	1970	27	73	45	55	33	77			58	42	49	51	29	71	24	76
		1979	37	63	54	46	39	61			66	34	52	48	35	65	28	72
Locus of European Social Innovations 1950–80	Non-Catering	1970	74	26	74	26	75	25			83	17	63	37	58	42	63	37
		1979	76	24	79	21	75	25			84	18	63	37	67	33	63	37
	Catering	1970	0	100	0	100	0	100			0	100	0	100	0	100	0	100
		1979	0	100	0	100	0	100			0	100	0	100	0	100	0	100
	E. Transport	1970	77	23	75	25	81	19	73	27	74	26	74	26	73	27	66	34
		1979	83	17	71	29	78	22	74	26	76	24	82	18	73	27	66	34
	F. Education	1970	–	–	0	100	0	100			0	100	0	100	0	100	0	100
		1979	–	–	0	100	0	100			0	100	0	100	0	100	0	100
Predominantly Public Provision	G. Medicine	1970	27	73	35	65	22	78			8	92	62	38	50	50	63	37
		1979	27	73	36	64	21	79			13	87	62	38	41	59	57	43

TABLE 6.5b *Change in modal split 1970–9*

(Shift to Goods)	B	D	F	G	H	Ir	H	UK	–	0	+
A. Food	0	0	0	–	0	0	0	0	0	7	0
B. Shelter	+11	+2	–1	–	0	–3	0	0	2	3	2
C. Domestic	+ 7	+8	+7	–	0	+4	+1	0	0	1	5
D. Entertain.	+10	+9	+6	–	+8	+3	+6	+4	0	0	7
inc. Non-Catering	+ 2	+5	0	–	+1	0	+9	0	0	3	4
Catering	0	0	0	–	0	0	0	0	0	7	0
E. Transport	+ 6	–4	–3	+1	+2	+8	0	0	2	2	4
F. Education	–	–	0	–	0	0	0	0	0	5	0
G. Medicine	0	+1	–1	–	+5	0	–9	–5	3	2	2

have been, until this decade, inappropriate for social innovation outside these areas.)

So, even in a period of relatively low economic growth, we can find evidence of both sorts of effect on patterns of final expenditure. What is the aggregate consequence of Engel's Law and social innovation on the distribution of final expenditure? Table 6.6, which is constructed by summing the categories in the columns of Table 1 (and also separating primary from manufactured products) shows a rather unexpected pattern of change over the 1970s. In four of the seven countries for which the data are available, final expenditure on marketed services *fell* as a proportion of all final expenditure, while the proportion devoted to manufactured products rose in six countries and fell in none. This is almost the reverse of the pattern of development that would be expected from the Engel's Law relationship; it results from the process of social innovation, the substitution of goods (and unpaid labour) for final services purchased from the formal economy.

This decline in final expenditure on marketed services leaves us with something of a problem. We know that employment in the service industries has been growing; Table 6.7[4] shows a regular pattern of growth of employment both in marketed and non-marketed service industries in Europe during the 1970s; similar patterns are to be found throughout the developed world from the early 1960s. How can we reconcile the discrepancies between the patterns of change in final expenditure and the industrial distribution of employment?

TABLE 6.6 *Distribution of all final consumption by commodity*

	BELGIUM			DENMARK			FRANCE			GERMANY		
	1970	1979	Change	1970	1979	Change	1970	1979	Change	1970	1979	Change
Primary	35.6	32.7	−2.9	23.0	29.3	−3.7	34.2	29.8	−4.4	(insufficient Data)		
Manufactures	30.1	23.0	+2.9	20.6	20.6	0	22.2	25.0	−2.8			
Marketed Services	21.6	20.9	−0.7	15.1	13.5	−1.6	21.8	24.5	+3.7			
Non-Marketed Services	22.7	23.4	+0.7	31.3	36.6	+5.3	21.8	20.7	−1.1			

	HOLLAND			IRELAND			ITALY			UK		
	1970	1979	Change	1970	1979	Change	1970	1979	Change	1970	1979	Change
Primary	32.5	29.7	−2.8	50.8	45.5	−5.3	44.4	41.6	−2.8	32.7	30.5	−2.2
Manufactures	20.4	23.2	+2.8	16.1	19.0	+2.9	15.9	18.3	+2.4	17.1	18.6	+1.5
Marketed Services	20.1	21.3	+1.2	13.6	12.4	−1.2	19.6	19.5	−0.1	22.1	22.2	+0.1
Non-Marketed Services	27.0	25.8	−1.2	19.5	23.1	+3.6	20.1	20.5	+0.4	28.1	28.8	+0.7

	row counts		
	—	0	+
Primary	7	0	0
Manufactures	0	0	0
Marketed Services	4	0	3
Non-marketed Services	2	0	5

TABLE 6.7 *Changing distribution of employment 1970–9* (%)

	BELGIUM			HOLLAND			FRANCE			GERMANY		
	1970	1979	Change	1970	1979	Change	1970	1979	Change	1970	1979	Change
1. Agriculture	4.9	3.3	−1.6	7.0	5.9	−1.6	13.2	9.0	−4.2	8.5	6.0	−2.5
2. Fuel and Power	2.1	1.6	−0.5	1.5	1.3	−0.2	1.6	1.4	−0.2	2.0	1.8	−0.2
3. Manufacturing	30.4	24.1	−6.3	25.4	20.4	−5.1	26.4	25.3	−1.1	36.8	33.6	−3.2
4. Construction	8.5	8.2	−0.3	10.8	9.7	−1.1	9.6	8.5	−1.1	8.1	7.5	−0.6
5. Market Services	35.8	42.4	+5.8	41.2	46.2	+5.0	32.8	37.9	+5.1	31.0	33.4	+2.4
6. Non-Market Services	17.3	20.4	+3.1	14.1	16.5	+2.4	16.5	18.0	+1.5	13.6	17.7	+4.1

	ITALY			UK			Row Counts		
	1970	1979	Change	1970	1979	Change	—	0	+
1. Agriculture	18.3	13.8	−4.5	2.8	2.6	−0.2	6	0	0
2. Fuel and Power	0.9	0.9	0	2.6	2.4	−0.2	5	1	0
3. Manufacturing	27.8	27.3	−0.5	30.3	27.6	−2.7	6	0	0
4. Construction	10.3	8.3	−2.0	7.3	6.9	−0.4	6	0	0
5. Market Services	28.6	32.2	+3.6	37.2	39.1	+1.9	0	0	6
6. Non-Market Services	14.2	17.5	+3.5	19.7	21.4	+1.7	0	0	6

There are, as we suggested in Section 6.1, two different sorts of possible explanation. The first rests on the observation that there is no one-to-one correspondence between the output of a particular industry and the final consumption of a particular sort of commodity. Simply, the commodities produced by some industries never reach final consumers directly, but are used as intermediate inputs to further production processes. Households seldom buy industrial lathes or semi-finished car bodies, even though these are the outputs of particular industries; rather, households buy the final commodities that they have been used to produce. And just as some manufacturing industries produce commodities which are not directly sold to final consumers, so some marketed service industries produce 'intermediate' outputs. Engineering consultancies and contract cleaning companies, for example, produce services, and are counted within the marketed services sector, yet sell their output predominantly to firms in the manufacturing sector rather than to households. We actually consume their products indirectly, as embodied in manufactured goods. Less obviously, we very seldom pay directly for the services produced by the retail distribution industry; the cost of sales services is bundled up in the price paid by the consumer for the manufactured commodity – again, though in a slightly different way these are intermediate services embodied in manufacture.

We can put this more concretely in terms of the simple input/output matrix in Table 6.8. In the UK in 1979, primary and manufacturing industries accounted for about 37 per cent of all value added (i.e. net output), marketed services accounted for about 48 per cent and non-marketed service industries about 15 per cent. But by contrast, about 50 per cent of all expenditure by households was on manufactures, and 22 per cent went to purchase marketed services. We can only account for the differences between the proportions by assuming the pattern of intermediate inputs from industries to final commodities is something like that suggested in Table 6.8. (The co-efficients inside the matrix in Table 6.8 are only estimates, since the published UK I/0 matrices treat investment goods conventionally as final output – the inter-industry flow of investment products is obscured.) While perhaps three-quarters of the net output of manufacturing and primary industry is

TABLE 6.8 *Simple input/output matrix for a closed economy*

Industries ↓ / *Commodities* →	Primary, Manufactures	Marketed Services	Non-Marketed Services	Total Value Added[1]
Primary, Manufacturing	27[3]	5	5	37
Marketed Services	23	17	8	48
Non-Marketed Services	0	0	15	15
Total Household Consumption[1,2]	50	22	28	100

NOTES

1 Total Value Added and Total Household Consumption are distributed in proportions as in the UK in 1979.

2 Total Household Consumption includes governmental direct (i.e. non-transfer payment) expenditure in the 'non-marketed' services category.

3 Co-efficients within the I/O matrix are hypothetical.

consumed as final manufactured commodities, with the remaining quarter going as intermediate inputs to service production, perhaps only one third of the total output of marketed service industries is finally embodied in marketed services directly purchased by households – and half of this total output may be embodied in goods.

So we have a category of marketed service industries – we shall refer to them collectively as 'producer services'[5] – which do not provide services for final consumption. These, clearly, provide one possible explanation for the discrepancy between the growth of marketed service consumption and employment; the growth in employment *could* be accounted for by a growth in the producer service industries. Table 6.9[6] shows that for those European countries for which statistics are available, value added by marketed service industries is about twice as

TABLE 6.9 *Distributions of value added compared with distributions of final consumption by branch*

	BELGIUM				FRANCE			
	Consumption		Value Added		Consumption		Value Added	
	1970	1979	1970	1979	1970	1979	1970	1979
Primary, Manufacturing, and Construction	55.7	55.7	46.7	47.0	56.4	54.8	49.5	44.9
Marketed Services	21.6	20.9	41.5	40.7	21.8	24.5	39.9	44.9
Non-Marketed Services	22.7	23.4	11.8	12.3	21.8	20.7	11.6	10.2

	HOLLAND				ITALY			
	Consumption		Value Added		Consumption		Value Added	
	1970	1979	1970	1979	1970	1979	1970	1979
Primary, Manufacturing, and Construction	52.9	52.9	47.3	45.7	60.3	58.9	50.7	49.0
Marketed Services	20.1	21.3	39.2	41.7	19.6	19.5	38.2	40.2
Non-Marketed Services	27.0	25.8	13.5	12.6	20.1	20.5	11.1	10.8

	UK			
	Consumption		Value Added	
	1970	1979	1970	1979
Primary, Manufacturing, and Construction	49.8	49.1	39.8	37.4
Marketed Services	22.1	22.2	46.8	47.7
Non-Marketed Services	28.1	28.8	13.4	14.9

large as final expenditure on final marketed services; the implication is that about half of the output of the marketed services sector consists of producer services. It is also clear from Table 6.9 that the producer service proportion of output from marketed services has, in most cases, been growing faster than the final service proportion: a small part of the discrepancy between final demand for services and employment in services is certainly explained in this way. But this is by no means the whole story.

The second sort of explanation that is available to us relates to the labour productivity levels of the individual industries. Obviously, if productivity in manufacturing industry is growing faster than productivity in service industry, then a rising proportional expenditure on goods is consistent with an increase in employment in services relative to employment in manufacturing industry. Table 6.10 demonstrates that labour productivity growth[7] in the services is indeed substantially lower than that in other sectors. So the contrast between the evolution of final demand for marketed services and the evolution of employment in marketed service industries is explained in part by the development of the producer services, and in part by the relatively low labour productivity growth rate in the service sector.

This 'productivity gap' between the services and manufacturing industry is quite crucial to our argument; it is one of the causes of social innovation. One of the main reasons that we drive private cars rather than paying for 'finished' transport services is simply that private motoring, on the whole, is cheaper. And one reason for this is precisely the productivity gap; motor manufacturers tend to become more efficient in their use of labour whereas public transport systems have in the past shown a tendency not to do so. In general, final service prices tend to rise relative to others because of service industries' low potential for innovation.[8] In other words, the effect of the productivity gap on employment works in two ways: it increases the employment generated by a given proportion of final expenditure on marketed services; but it decreases that expenditure in so far as it encourages social innovation.

We might speculate that a rather similar process applies to non-marketed services. These also show relatively low labour

TABLE 6.10 *Productivity growth by sector as proportion of national average*

	BELGIUM	HOLLAND	FRANCE	GERMANY	ITALY	UK
1. Agriculture	1.18	1.28	1.12	1.24	1.14	1.13
2. Fuel and Power	1.59	1.79	1.15	1.09	0.94	1.06
3. Manufacturing	1.30	1.18	1.03	1.04	1.09	1.04
4. Construction	0.93	0.78	0.88	1.02	0.92	0.94
5. Market Services	0.86	0.95	0.97	1.00	0.93	0.97
6. Non-Market Services	0.88	0.81	0.80	0.79	0.80	1.04
National Average, 1970s	1.32		1.36	1.35	1.25	1.02

	Below Average	Average	Above Average
1. Agriculture	0	0	0
2. Fuel and Power	1	0	5
3. Manufacturing	0	0	6
4. Construction	5	0	1
5. Market and Services	5	1	0
6. Non-Market Services	5	0	1

productivity growth; though we do not pay directly for each item, we do pay indirectly for the whole basket of public services through our taxes. Low productivity growth in non-marketed services means that extra taxes go disproportionately to pay for high real wages for unchanged jobs (i.e. maintaining wages relative to similar occupations in industries with higher productivity growth). So given levels of non-marketed service provision become increasingly expensive as time passes. It could be that these continuously rising effective prices of public services are part of the explanation for the increasing unpopularity of welfare spending. Social innovations that enable alternative modes of provision for the sorts of services produced on a collective basis may have been, in the past, difficult to find (though, as we shall see, this situation may change in the future). But nevertheless, to put the matter at its bluntest, the fact that increments of expenditure on non-marketed services are as likely to go to increase service workers' real wages without increasing service output as to raise the level of service provision, must act as something of a disincentive to the growth of the non-market service sector. (We should, however, bear in mind that whatever the real facts about the evolution of output and productivity in the public service sector, the 'tax revolts' which have, in one form or another, been a ubiquitous phenomenon throughout the developed world, are orchestrated in the main by politicians ill-disposed towards the welfare state. So it would perhaps be appropriately cautious to give this particular part of the argument the status of an unproved hypothesis.)

In general, however, the arguments in this Section do seem an adequate basis for a plausible general description of the linkages between Engel's Law, social innovation, and the structure of employment. But this only gets us part way through the model. In the next chapter we will consider the remaining element, the *occupational* structure of employment.

NOTES

[1] Descriptions of the derivation of these statistics will be found in footnotes to Section 4.3.

[2] This data is calculated from *National Accounts, ESA: Detailed Tables by Branch 1970–1979*, EEC. Brussels, 1981, Tables 5; and *General*

Government Accounts and Statistics 1971–1978, EEC Brussels, 1981, expenditures are deflated by the appropriate price indices provided by the ESA, and expressed as proportions of total household and direct (i.e. non-transfer-payment) government expenditure.

[3] Sources; as in n. 2, *excluding* government expenditure.

[4] Source; *National Accounts, ESA; Detailed Tables by Branch*, Table 4.

[5] This term was originated by H. I. Greenfield in *Manpower and the Growth of Producer Services*, Columbia University Press, New York, 1966.

[6] Source; *National Accounts, ESA; Detailed Tables by Branch*, Table 2.

[7] Calculated here by dividing each sectors change in value added at constant prices by change in numbers of employees.

[8] This proposition is demonstrated by J. Skolka 'Long Term Effects of Unbalanced Labour Productivity Growth' in Solari and Pasquier (eds.), *Private and Enlarged Consumption*, North Holland, 1976.

CHAPTER 7

The National Accounting Model 2: From Industrial to Occupational Employment

7.1 Industries and Occupations

LET us return for a moment to some imaginary dawn of economic organization. The categories of human activity that we would think of as 'economic' consist of hunting for meat, gathering plants, hewing wood for fires, drawing water. There are no industries, no occupations in the modern sense of the word. Certainly people are *occupied* in some activities; but, assuming perhaps a certain rudimentary sexual division of labour, we would not consider describing any particular man as 'a hunter by occupation' – because *every* man hunts. Similarly we do not think of the 'gathering industry' because, in our ideal prehistoric commonwealth, *every* woman devotes some of her time to this. These categorizations of economic activity are only useful when they enable us to distinguish between different sorts of people.

So far, in the book, we have demonstrated the complexity of the relationship between household behaviour and the structure of the developed economy. A particular pattern of household needs does not have any necessary ahistorical reflection in the distribution of output or employment of the various sectors of the economy. In this chapter we shall take the argument one step further. Just as household needs do not uniquely determine the industrial structure, so the industrial structure does not uniquely determine the pattern of jobs in the economy. Indeed, it will emerge that the organization of production within industries has a much greater effect on the occupational structure of employment (and hence unemployment) than does change in the distribution of demand among the products of the various industries. But let us remain for the moment in the Stone Age. How do distinct industries and occupations emerge from the initial state of undifferentiated subsistence activities? Why do individuals specialize in particular activities – and what is the relationship between specialized occupations and the industrial structures in which they are employed?

With no trade, the primitive group must acquire all its subsistence requirements itself, so economic activities are undifferentiated. All groups must carry out all subsistence activities. Within the groups there may be some specialization, by the stage in the life cycle or by sex, but even this rudimentary division of labour would be upset by such small scale demographic crises as an imbalance between the sexes or a premature death. The development of particular technologies might serve to intensify the division of labour within the groups, but still without trade, the group as a whole is left to produce all of its own necessities. However, once the group begins to trade some particular commodity with other groups, our normal categorizations of economic activity begin to be useful.

As family groups or communities begin to specialize in the production of particular commodities over and above their own current needs, we see the beginnings of a familiar economic structure. There is still a wide range of activities in the undifferentiated sphere; though some commodities are traded, still most economic activities are here – each primary group still produces *all* its requirements for *most* commodities (just as all modern households produce some commodities – cooking and cleaning – in the undifferentiated, 'informal' sphere). But nevertheless there is now some sense in classifying individuals (or more likely, families or communities) by those particular production activities that they undertake in the differentiated sphere, producing commodities for trade. That is to say that, while all households perhaps still forage for edible vegetation, and some make pots for their own use, some groups can be meaningfully described as potters in that they make pots for the use of other groups (just as we would now describe some people as cooks or carpenters in spite of the fact that, outside the differentiated sphere of production for exchange, the carpenters, in common with everyone else in the society,[1] also cook meals for their own and their household's consumption).

At one stage of this idealized economic history, we can think of 'occupations' and 'industries' as being essentially identical. Some special knowledge, or privileged geographical position, or suitable form of social organization, gives an advantage to an individual group in the production of a particular commodity. If each individual carries out all of the tasks involved in the production of a particular traded commodity then we would see the individual as the possessor of an 'occupation' in the modern sense of the word – and the set of individuals with that one occupation would comprise an 'industry'. If the potter carries out all the tasks – digging clay, kneading, forming, painting, firing, and trading – necessary for the production of pots, then employment in the neolithic domestic utensils industry consists entirely of potters.

At another stage, however, another set of advantages may lead to the division of the various tasks involved in the production of a particular commodity. These advantages are the ones identified by Adam Smith; increasing efficiency from specialization in a less complex task, the possibility of using special equipment efficiently, and the effect of specialization in the encouragement of further innovation. And in addition, the possessors of special knowledge or expertise can improve their positions by employing less costly labour for the less skilled operations. Now each industry may employ a number of different occupations. The economic unit that produces pots now employs labourers to dig clay, skilled workers to form and decorate, linguists, perhaps, to travel with the wares to distant parts.

Before we get too taken up with this exercise of imagination, we should remember that it is not intended as a genuine reconstruction of the development of the division of labour. The imaginary neolithic has been introduced, not to assert any particular view of economic history, but to point the distinction between a number of economic categories. We have undifferentiated production of commodities within households, by household members, for their own households' consumption – in prehistory including all or most production activities, in modern societies including those activities which we have classified being in the 'informal economy' (though excluding 'hidden' production). Distinct from these undifferentiated activities, we have 'occupations', categories describing sets of activities carried out by individuals, contributing to the production of commodities which are traded outside those individuals' households. These occupations may comprise the complete set of activities necessary to produce a particular commodity, but more usually groups of occupations cooperate in the production of commodities – such groups are 'industries'.

This set of distinctions is interesting to us because the boundaries change. We have considered, in previous chapters, some of the ways in which particular activities shift between the undifferentiated informal sphere and the formal economy. In this chapter we shall consider some of the ways in which there may be changes in the set of activities which constitute an occupation, and in the distribution of occupations within industries.

7.2 Three Processes of Change

It will help us to visualize the distribution of employment in the formal economy in a two dimensional matrix. On the horizontal axis we have the range of occupations, going from, perhaps, administrative and professional workers on the left, through all the service occupations, clerical, sales, catering, to the manual occupations on the right. On the vertical axis we have the set of industries, from primary industries, farming, mining, at the top, through manufacturing industry to the service industries at the bottom. Each worker in the formal economy falls into one of the categories in each dimension. Each worker has an occupation, which he or she employs in the production of some commodity; the nature of this commodity identifies the worker's industry. So each job can be classified as lying within one cell of the employment matrix.[2]

In the simplest of our imaginary prehistorical cases, each industry consisted of just one occupation; in this case the matrix would look rather sparse, with all the entries on each horizontal line concentrated in one cell. (So in the horizontal row associated with the pottery industry, there would be just one cell – that lying in the occupational column for potters – with entries, while all the other cells, those associated with clerks, or sales staff, or unskilled labourers, would be empty.)

As the Adam Smith advantages develop, so the occupational concentrations within industries start to dissipate. Industrial processes change, the set of tasks contained within each occupational category is reduced in number, and within each industry an ever wider range of ever more specialized occupations develops. The clerical functions, the management and marketing and manual labouring that were once all circumscribed within the old 'potter' occupation are now carried out by specialized clerical and management and labouring occupations. At the most extreme point in the division of labour, and as the scale of enterprises grow, we come to the point that each industry employs some members of every specialized occupation – not just clerks and labourers, but nurses to attend to accidents in the kilns, and teachers and gardeners. From a sparsely populated matrix with only one cell entry in each row, we move to a much more well distributed matrix with entries in every cell.

But another sort of change acts against the influences of the Adam Smith advantages. In the previous chapter we discussed the situation in which particular industries, instead of carrying out all the processes necessary for the production of some final commodity, only carry out some of them, subcontracting the rest to other industries (or act as subcontractors). Some firms produce investment goods which are in effect intermediate inputs to further production processes, or semi-finished 'intermediate' goods, or 'producer services'. Frequently the products of these 'intermediate' industries are purchased by firms as an alternative to employing people in equivalent occupations within the firm. This pattern (though relevant to manufacturing occupations) is most clearly visible in relation to service occupations; the pottery, rather than employing marketing executives, might alternatively employ the services of a marketing agency.

There are a number of reasons for this sort of sub-contracting. The most obvious is simply the pressure of economies of scale. Where some particular component, or piece of productive machinery, constitutes only a very small part of a firm's output, and where the same component or device is used throughout the industry, a firm which specializes in the production of that component or device on a subcontracting basis can often achieve relative advantages of scale. Similarly, services, which might be provided on a small scale and relatively inefficiently by the firms which use them, may be provided much more efficiently using more costly capital equipment, by a specialist subcontracting service agency. The economies of scale go further than the mere availability of capital equipment, however. Specialist service firms in particular can have more effective supervision and management of particular service functions than could other firms employing direct labour for the same task. So, for example, a cleaner employed directly by a manufacturing firm might not be subject to any very clear management control, where a subcontracting industrial cleaning firm would be able to supervise the task quite effectively. In addition, subcontracting rather than direct employment has other advantages; it may reduce overheads such as office costs; it reduces liability for redundancy payments where employment is unstable – it may give the firm access to cheaper (i.e. non-unionized) labour; and there may also be fiscal advantages (e.g. to take advantages of government subsidies to small firms), or reasons related to anti-trust or tax legislation.

Just as the Adam Smith pressures tend to disperse and divide occupational employment, from the central occupation in each industry to a wide range of more specialized occupations each with a narrower set of tasks, so the subcontracting pressures tend to a reconcentration of occupations. The division of labour within industries increases the dispersion along the industrial rows of our employment matrix, while the subcontracting of occupational functions leads to a concentration within particular cells in the occupational columns; in effect, we move away from the prehistoric situation of one cell-entry per row, towards the post-industrial cottage economy of one cell-entry per column.

So we can identify three separate influences on the distribution of employment across industries and occupations. First are the changes discussed in previous chapters concerning the pattern of final household demand for commodities – the Engel's Law and social innovation effects. Second are the effect of the changing division of labour within particular industries. Third are the consequences of the changing pattern of subcontracting and intermediate demand between industries.

7.3 Occupational Change Within Industries

So far we have proceeded entirely by way of introspective analysis; to get any further we must try to take a more empirical view. But empirical analysis is not very easy in this field. While statistical authorities in most countries collect data on employment as classified by industry on a routine basis, they are less likely to do so for employment classified by occupation, and even less likely to do so for employment cross-classified by employment and occupation. There is a perfectly natural explanation for this. The industrial classification is easy to compile. An industry consists of a set of firms which produce the same commodity, and all the employees of a firm in an industry are employed in that industry, so the data is simple to collect and classify. The occupational classification is much more problematical. The occupational categories themselves must be defined unambiguously – and since each occupation consists of a bundle of tasks, which may vary between firms, and between industries, and over time this is much less easy to draw up than the equivalent classification of industries by the major products of firms. And each *individual* (as opposed to each *firm* in the industrial classification) must be allotted to a particular category.

So while industrial employment data are, throughout the developed world, regularly published on a monthly or annual basis, occupational data are likely to be produced only on a ten-yearly (i.e. census) basis, or not at all. Where occupational information is collected more regularly, it tends to be on a partial basis (e.g. for a part of manufacturing industry only), or to use very aggregated occupational categories (e.g. manual/non-manual), or from a survey whose sample size precludes any very substantial disaggregation of either the industrial or the occupational categories. Finding out what has been happening to the industrial/occupational distribution of employment in the recent past is very difficult.

There are some data, however, and by combining various sources we can often construct a reasonably reliable picture. In the following pages we shall consider changes in the UK occupational distribution during the 1970s. (The estimates have been derived by combining information on the occupational distribution of employment within industries from the New Earnings Survey with the annual industrial Census of Employment; we then compare five-year averages to smooth sampling fluctuations. More information about this process, together with detailed data, and a selection of comparable

TABLE 7.1 *Summary occupational/industrial distribution of employment UK (% of total employment)*

Occupations

1968–73

Industries	Administrative, Professional, Technical, Clerical	Other White Collar	Manual	All
Primary	0.4	0.1	3.4	3.9
Manufacturing	8.1	2.1	25.7	35.9
Utilities	1.6	0.3	5.8	7.7
Services	24.4	15.9	12.2	52.5
All Sectors	34.5	18.4	47.1	100.0

1974–8

	Administrative, Professional, Technical, Clerical	Other White Collar	Manual	All
Primary	0.5	0.1	2.7	3.3
Manufacturing	8.2	1.9	22.5	32.6
Utilities	2.1	0.2	4.9	7.2
Services	31.1	16.2	9.6	56.9
All Sectors	41.9	18.3	39.8	100.0

CHANGE 1968–73 TO 1974–8

	Administrative, Professional, Technical, Clerical	Other White Collar	Manual	All
Primary	0.1	0.0	−0.6	−0.6
Manufacturing	0.2	−0.2	−3.3	−3.3
Utilities	0.4	0.0	−0.9	−0.5
Services	6.7	0.3	−2.6	4.4
All Sectors	7.4	0.1	−7.4	

material from other European countries, will be found in the Appendix to this Chapter.)

Table 7.1 gives the most aggregated picture of the change in the occupational/industrial employment distribution. We have, as in the conceptual matrix described in the previous section, occupational categories in the columns of the table. In this case we divide occupations into just three categories – administrative, technical, professional and clerical workers in the first group, other white collar and service occupations (including sales, catering cleaning, security, and transport) in the second, and all remaining manual occupations in the third. Again, in the rows of the table we have industrial categories, in this case primary (farming and mining), manufacturing, utilities (electricity, gas, water, and construction), and services. The entries in each cell give the percentage of the working population falling into each combination of industry and occupation (or, in the third part of the table, the change in this proportion between the earlier and the later part of the 1970s).

Looking first at the change in the totals in employment in the industrial categories, we see the pattern that we would expect from Chapter 6. Employment in those industrial sectors directly concerned with material production – primary, manufacturing, and utilities – has in each case fallen as a proportion of total employment, while the proportion in the service sector has grown. The data in the Appendix to this chapter shows an equivalent pattern in all the other European countries – and for the 1970s at least, it clearly holds for the whole of the developed world. Total employment in the three occupational classes shows a slightly less expected pattern of change, however. Certainly, as we would expect, the proportion of total employment in manual occupations has fallen quite considerably over the relatively short period – by a little less than 7½ per cent. But this proportional fall in manual employment has not been to the benefit of all the categories of service workers. All the increase is concentrated in the first of the categories, among the professional, technical, administrative, and clerical workers – what we might think of as the information or knowledge workers – whereas the remaining service categories show no proportional increase. Within the individual occupational categories for each industry, the change has been quite regular; the manual employees in all sectors have declined as a proportion of the total, while the 'knowledge' workers in each sector have increased – and a small decline in the other service category in manufacturing industry has been compensated for by an equivalent small increase in the service industries.

Do these developments reflect changes in the pattern of occupational employment within industrial sectors or changes in the pattern of demand for the products of those sectors? We can begin to answer this question by looking at the distribution of employment between the occupations in each individual sector. Table 7.2 gives a rather disaggregated view. It expresses the employment in each industrial/occupational category as a proportion of the total in the particular industrial sector (i.e. 'row percentages').

The bottom row of this table gives us the same message, though in a more disaggregated form, as the bottom row of table 7.1. We see that the first three occupational categories, administrative, technical, professional, and clerical workers have all increased as a proportion of total employment. We see that while manual workers considered on the broadest possible basis constitute a declining proportion of the workforce, transport and communications workers have increased their proportion somewhat. And the unchanged proportion in the broad 'other white collar' category hides a decline in the number of sales workers compensated for by an increase in the numbers of security, catering, and cleaning workers.

But the more important messages of table 7.2 are to be found in the other rows. Consider the proportions in the more narrowly defined 'other manual' occupational category. In all cases the proportion declined; here we must have the operation of the traditional Adam Smith processes of division of labour. Production becomes more mechanized, and capital intensive, more demanding of more specialized administrators. As production becomes more reliant on advanced techniques, so more technical workers are needed; the growth of information processing required by the larger scale and the increasing complication of production implies an increase in the need for clerical and other information handling and transmitting services. And the need for unspecialized manual workers proportionately declines. The column of negative values for

TABLE 7.2 *Occupational distributions within industrial sectors, UK (% of employment in industry)*

Occupations → *Industries ↓*	Administrative and Technical	Educational Medical Professions	Clerical	Sales Workers	Security Workers	Catering and Cleaning	Farming Gardening	Transport and Communications Workers	Other Manual	All
ROW PER CENT, 1968–73										
PRIMARY	4.9	0.1	5.4	0.9	0.1	1.7	37.2	4.9	44.4	100.0
Manufacturing	10.8	0.3	11.4	2.9	0.4	3.5	0.1	3.0	68.4	100.0
Utilities	11.3	0.1	9.8	1.3	0.3	1.8	0.2	3.9	71.5	100.0
All Services	11.3	12.5	22.6	11.4	1.8	17.1	1.0	7.1	15.2	100.0
Trans/Telecoms	8.4	0.2	29.5	0.8	0.7	3.8	0.1	35.6	20.9	100.0
Distribution	11.3	0.2	16.6	41.9	0.2	4.9	0.3	5.5	19.9	100.0
Ins/Banking	16.1	0.2	60.5	9.0	0.7	7.1	0.9	0.5	5.0	100.0
Prof/Ed. Servs.	9.9	45.0	12.3	0.2	0.9	25.9	0.9	0.5	4.5	100.0
Misc. Serv.	10.0	3.3	15.9	6.3	0.6	38.3	1.1	2.4	22.0	100.0
Pub. Admin.	15.4	3.0	31.2	0.0	10.2	13.5	3.0	3.6	20.0	100.0
All Sectors	10.8	6.7	16.9	7.1	1.2	10.1	2.0	5.3	39.8	100.0
ROW PER CENT, 1974–8										
Primary	8.2	0.3	6.2	0.7	0.2	1.4	37.9	8.5	36.5	100.0
Manufacturing	13.0	0.4	11.9	2.4	0.5	2.8	0.1	7.2	61.7	100.0
Utilities	16.4	0.3	11.9	1.2	0.4	1.4	0.2	7.1	61.2	100.0
All Services	14.4	15.5	24.9	9.4	2.1	16.9	0.9	7.1	8.9	100.0

	Administrative and Technical	Educational Medical Professions	Clerical	Sales Workers	Security Workers	Catering and Cleaning	Farming Gardening	Transport and Communications Workers	Other Manual	All
Trans/Telecoms	13.0	0.4	30.7	0.8	0.9	6.7	0.0	28.9	18.5	100.0
Distribution	16.3	0.7	20.2	36.1	0.3	3.6	0.2	10.3	12.2	100.0
Ins/Banking	21.8	0.5	60.6	7.6	1.5	4.0	0.4	1.0	2.6	100.0
Prof/Ed. Servs.	9.6	48.4	13.8	0.1	0.3	23.8	0.7	0.6	2.7	100.0
Misc. Servs.	12.5	8.1	15.5	4.6	0.6	40.2	1.2	3.5	13.6	100.0
Pub. Admin.	20.2	2.5	40.0	0.0	13.3	9.8	2.9	5.6	6.0	100.0
All Sectors	13.8	9.0	19.1	6.2	1.4	10.7	1.8	7.2	30.8	100.0
ROW PER CENT, CHANGE, 1968–73 to 1974–8										
Primary	3.3	0.1	0.8	−0.2	0.1	−0.3	0.7	3.6	−8.3	0.0
Manufacturing	2.2	0.2	0.4	−0.5	0.1	0.3	0.0	4.2	−6.7	0.0
Utilities	5.1	0.2	2.2	−0.1	0.0	−0.2	0.0	3.2	−10.3	0.0
All Services	3.1	2.9	2.2	−2.0	0.3	−0.2	−0.1	0.1	−6.3	0.0
Trans/Telecoms	4.6	0.2	1.2	0.0	0.2	2.9	0.0	−6.7	−2.3	0.0
Distribution	5.0	0.5	3.6	−4.9	0.1	−1.4	−0.1	4.8	−7.7	0.0
Ins/Banking	5.7	0.3	0.0	−1.3	0.8	−3.1	−0.5	0.5	−2.4	0.0
Prof/Ed. Servs.	−0.3	3.4	1.5	−0.1	−0.6	−2.1	−0.2	0.2	−1.8	0.0
Misc. Serv.	2.6	4.8	−0.4	−1.6	0.0	1.9	0.2	1.1	−8.4	0.0
Pub. Admin.	4.8	−0.6	8.8	0.0	3.1	−3.7	−0.1	1.9	−14.1	0.0
All Sectors	3.0	2.3	2.1	−0.9	0.3	0.6	−0.2	1.9	−9.0	0.0

the manual occupations in the third section of this table reflects this most obvious consequence of process innovation within industries.

Balancing these declining proportions are the increases in the 'knowledge worker' categories. In the 'administrative, technical' column, the only industry showing a decline in the proportion of these occupations is professional and educational services (consisting mostly of medicine and education, though with some technical producer services as well). The declining proportion here does not imply any reduction in absolute numbers; on the contrary, the number of administrators and technical specialists in this industrial sector has increased. Rather it reflects a change in the balance of demand within the sector, with education and medicine (which have a very small administrative and technical 'overhead' and naturally a very large proportion of their employment in the educational and medical professions) growing proportionately faster than the other industries within this sector. Similarly the declining proportion of education and medical workers in public administration and the declining proportion of clerical workers in 'miscellaneous services' reflects the changing compositions of the branches within these sectors, and not an absolute decline in the numbers of these occupations in any one branch. And with these exceptions the knowledge workers are increasing their proportion of employment in every industrial sector.

Within this group of knowledge workers, however, there is a regular pattern of change in the balance between the secretarial category and the administrative, technical, and professional groups. In most cases (that is, with the exception of the professional and educational sector, and public administration), the ratio of clerical workers to the other knowledge worker categories has been declining. In other words, over this period, the average clerical worker has been able to 'service' an increasing number of administrative technical and professional workers.[3] In manufacturing industry for instance, in the early 1970s, there was one secretarial worker for every .97 of an administrative, professional or technical worker, where, by the later part of the decade, there was one clerical worker to every 1.13 other knowledge worker. Similarly, in the services, a ratio of one clerical worker to every 1.05 administrative technical or professional worker in the earlier period, was succeeded by a ratio of 1:1.20 in the later. The reason for this change in ratios – which is to be found very generally across the British economy, and also holds in all the other major European economies – is probably an increase in efficiency in the provision of clerical services. When we consider the data broken down by sex it emerges that the ratio changes faster for men, whose employment in clerical functions has been concentrated in the insurance and other financial industries – industries where the introduction of computers has had major effects on clerical employment.[4] In general it seems that the change reflects both organizational innovations, such as the typing pool, and technical equipment (though the effect of the word-processor will not be visible in these data).

These then are two relatively straightforward consequences of process innovation: the proportion of manual workers declines relative to 'knowledge workers'; and, within the 'knowledge workers' category, clerical workers' proportions declining relative to others'. In both cases there seems to be clear set of technical and organizational reasons for the proportion employed in particular occupations to change in particular ways which apply to each of the industrial sectors in our table (and also for a much longer and more disaggregated list of industries). Some of the other categories of occupation appear from our table to have similarly simple patterns of evolution. Security workers seem to have increased as a proportion of employment in each sector other than professional and educational services; this presumably reflects changing social conditions rather than technology. The 'farming' occupation (which includes gardeners) is well distributed across the service sector, and while, on the basis of our estimates it appears that the absolute number of such service sectors 'farmers' has increased, it has risen rather more slowly than the general rate of increase in the number of service workers, their proportion of employment in these industries has fallen. The remaining occupations, however, have a rather more confusing and contradictory pattern of change.

Consider the transport and communications occupations. Overall there has been a quite substantial increase in the total employment in this category (an increase of 1.9 per cent of total employment). Transport and communications workers have increased as a proportion of total employment in each of the sectors in table 7.2, with one exception – the transport and telecommunications industry itself. This looks to be the precise converse to the 'subcontracting' process discussed in the previous section. Rather than purchasing intermediate services from specialized transport service firms, firms appear increasingly to provide transport services themselves. In principle this could be related to techno-economic pressures which reduce the costs of firms maintaining private transport fleets (such as a reduction in the capital costs of transport equipment). But in fact in this case it may be related to two circumstantial factors: the denationalization of the publicly owned monopoly freight corporation in the mid 1970s, and perhaps to tax advantages to corporations owning and operating their own transport fleets.

The final two occupational categories show some of the characteristics of subcontracting particular occupational functions. We can see the changes most clearly by presenting the data in the form of Table 7.3 (which returns to the percentage of total employment in the economy, 'table percentage' basis of Table 7.1 rather than the 'proportion of employment in the sector', row percentage basis of Tables 7.2). Employment in the catering and cleaning occupations is declining, and indeed, if we exclude for a moment employment in the miscellaneous services sector that provides these services on a subcontracting basis, then overall employment in these occupations would be seen to decline slightly. However, more than compensating for the decline in the direct employment of catering and cleaning workers in particular industries, has been a growth of such employment in the industry that provides the services on a contract basis.

The case of sales workers is rather more complicated. Here, as we see from Table 7.3, employment has fallen as a proportion of the total (or at least remained unchanged) in every sector, so that sales workers overall have declined from 7.1 to 6.2 per cent of total employment. And altogether the largest single component of this overall decline is the decline in the number of sales workers employed in the distribution industry itself – a decline equivalent to about half of one per cent of the total employment in the economy. And yet the distribution industry itself has grown somewhat in employment terms, by around 0.2 per cent of total employment.

So we have a combination of two different sorts of effect.

On the one hand, given that the proportion of all employment in the distribution industry has grown, we must conclude that other industries are tending on balance to rely increasingly on using its services rather than employing their own sales staffs directly. On the other hand, within the distributive industry itself, we have process innovation, reducing the proportion of sales workers in favour of the technical, administrative and other information workers necessary to run a more automated distributive system. In short, we have both subcontracting and the conventional division of labour.

7.4 Occupational vs. Industrial Change

Remember that our ultimate purpose is to develop a picture of the relationship between the pattern of final demand in the economy, and the pattern of employment and unemployment. There is now at least some evidence of change in the occupational balance of employment within industries. But how important is this sort of change? How large is its effect on change in the ultimate pattern of employment, relative to the changes in the demand factors discussed in the previous chapters? We cannot, using the industrial/occupational distribution alone distinguish between the effects of final demand changes and intermediate (i.e. subcontracting) demand changes – but, using a very simple 'shift-share' analysis, we can distinguish between the occupational effects and the joint consequences of two sorts of demand change.[5]

Consider Table 7.1: it gives us the actual occupation/industry distribution for the early and the late 1970s. It also provides us with a basis for speculation: what if total employment in each industrial sector had changed exactly as it actually did between 1968–73 and 1974–8, *but the occupational distribution stayed exactly as it was in 1968–73*? If we multiply the 1974–8 industrial distribution of employment by the 1968–73 occupational proportions in each industry from Table 7.1 we arrive at a synthetic estimate of the occupation/industrial employment pattern in the later period, assuming no within-industry occupational change. From this artificial construct we can estimate the effect of the change in the balance of demand between the industrial sectors – by subtracting the actual 1968–73 data from the synthetic 1974–8 data – and then estimate the effect of the changed occupational pattern

TABLE 7.3 *Occupational versus industrial employment: the sales and catering/cleaning examples*

Industries ↓		Admin., Prof., Tech., Clerical	Sales	Security	Catering, Cleaning	Other Manual	All Occupations
		← *Occupations* →					
Primary	– 68–73		0.0		0.1		3.9
	– 74–8		0.0		0.0		3.3
	– change		0.0		0.0		0.6
Manufacturing	– 68–73		1.0		0.9		35.9
	– 74–8		0.3		0.9		32.6
	– change		–0.3		0.0		–3.3
Utilities	– 68–73		0.1		0.1		7.7
	– 74–8		0.1		0.1		7.2
	– change		0.0		0.0		–0.5
Transport etc.	– 68–73		0.1		0.3		7.0
	– 74–8		0.1		0.4		6.6
	– change		0.0		+0.2		–0.4
Distribution	– 68–73	3.4	4.9	0.0	0.6	3.1	12.0
	– 74–8	4.6	4.4	0.0	0.4	2.8	12.2
	– change	+1.2	–0.5	0.0	–0.2	–0.3	+0.2

Financial etc.	– 68–73		0.4		0.3		4.4
	– 74–8		0.4		0.3		5.0
	– change		0.0		−0.1		+0.7
Personal etc.	– 68–73	11.5	0.6	0.2	7.0	3.1	22.4
	– 74–8	15.1	0.5	0.1	7.9	2.5	27.0
	– change	+3.6	−0.1	−0.1	+0.9	−0.9	+3.6
Public Administration	– 68–73		0.0		0.9		6.8
	– 74–8		0.0		0.7		7.1
	– change		0.0		−0.2		+0.3
All Industries	– 68–73	34.4	7.1	1.2	10.1	47.1	100.0
	– 74–8	41.9	6.2	1.4	10.7	39.8	100.0
	– change	+7.4	−0.9	+0.3	+0.6	−7.3	0.0

TABLE 7.4 *Shift-share analysis of industrial/occupational employment distributions*

Industries ↓	Administrative Professional Technical Clerical	*Occupations* → Other White Collar	Manual	All
1. Estimated Employment Distribution 1974–8, No Occupational change				
Primary	0.4	0.1	2.9	3.3
Manufacturing	7.3	1.8	23.3	32.6
Utilities	1.5	0.2	5.4	7.2
Services	26.9	17.4	12.6	56.9
All Sectors	36.1	19.6	44.2	100.0
2. Sectoral Demand Effects				
Primary	0.0	0.0	−0.5	−0.6
Manufacturing	−0.8	−0.2	−2.4	−3.3
Utilities	−0.1	0.0	−0.5	−0.5
Services	+2.5	+1.5	+0.4	+4.4
All Sectors	+1.7	+1.2	−2.9	0.0
3. Occupational Effects				
Primary	+0.1	0.0	−0.1	0.0
Manufacturing	+0.9	−0.1	−0.8	0.0
Utilities	+0.6	0.0	−0.5	0.0
Services	+4.3	−1.3	−3.1	0.0
All Sectors	+5.8	−1.3	−4.5	0.0
4. Summary				
All Sectors, 1968–74	34.5	18.4	47.1	100.0
+ Sectoral Demand Effect	+1.7	+1.2	−2.9	0.0
Estimate 1974–9 No Occupational Change	36.1	19.6	44.2	100.0
+ Occupational Effect	+5.8	−1.3	−4.5	0.0
Actual, 1974–78	41.9	18.3	39.8	100.0

within industries by subtracting the synthetic 1974–8 estimates from the actual 1974–8 data.

Table 7.4 presents such an exercise. From the summary section of the table it is immediately obvious that of the two components in the overall change in the balance between the broad occupational categories, the effect of occupational change within industries is the larger. Of the 7.5 per cent increase in the number of knowledge workers over the period, less than a quarter appears to be due to the growth in the service industries, while more than three-quarters comes from the increase in the proportion of employees in service occupations within industries. Similarly, less than one third of the decline in the proportion of employees in the broadly defined manual occupations comes from the decline in total employment in those industries with a high proportion of manual employees. And in the case of the 'other white collar' occupations, the growth in the overall size of the industries in which they are concentrated would lead us to expect an absolute increase in the proportion of employment – which means that, since the overall proportion of these employees has actually remained unchanged, the occupational effect is in fact negative.

Table 7.5 summarizes the results of a similar exercise carried out for the more disaggregated occupational categories. In only two of the nine occupational classes do the 'sectoral' effects outweight the 'occupational'. In farming, the main change comes from the contraction of the primary sector, with changes in occupational distributions having no significant effect. In the catering and cleaning occupations, the main growth comes from the increase in the size of the 'miscellaneous services' industrial sector; the 'expected' growth here is higher than that which actually occurred, implying that the growth industries showed on balance a tendency to reduce their catering and cleaning proportions (which we know to be true from Table 7.2). In all the remaining cases, the effects attributable to changing occupational distributions within industries comfortably outweigh the changes that we would expect on the basis of the changing balance of employment between industries.

The European data in the Appendix to this chapter also show, for rather aggregated industrial and occupational sectors for both the 1960s and the 1970s, an identical pattern of

TABLE 7.5 *Shift share summary, more detailed occupational categories*

	Administrative and Technical	Educational Medical Professions	Clerical	Sales Workers	Security Workers	Catering and Cleaning	Farming Gardening	Transport and Communications Workers	Other Manual	All
Actual Distribution 1968/73	10.8	6.7	16.9	7.1	1.2	10.1	2.0	5.3	39.8	100.0
+ Sectoral Demand effects	+0.1	+1.1	+0.5	+0.1	0.0	+1.1	−0.2	−0.2	−2.5	0.0
Estimated 1974/8 no occupational change	10.9	7.8	17.4	7.2	1.2	11.2	1.8	5.0	37.4	100.0
+ Occupational change effects	+2.9	+1.2	+1.7	−1.0	+0.2	−0.5	0.0	2.2	−6.6	0.0
Actual Distribution 1974/8	13.9	9.0	19.1	6.8	1.4	10.7	1.8	7.2	30.8	100.0

change; the occupational effect outweighing the sectoral effects by more than two to one. It seems reasonable to conclude, on the basis of this evidence, that the pattern of occupational employment in the economy as a whole is more affected by technical and organizational change in production processes than by change in the pattern of demand. *Irrespective* of the state of demand for intermediate or final services (i.e. the *output* of service industries), the development of improved production processes will lead to the increase in the demand for employment of workers in service (i.e. non-manual – and increasingly technical or managerial) occupations.

7.5 Occupations and Unemployment

We have one further, and final, link to establish in our chain of relationships which now connects household and collective provision for particular 'final functions' at one extreme, to the distribution of employment in various occupations at the other extreme. Even though employment statistics by occupation are very rarely collected or published, statistics on unemployment by occupation are quite readily available.

Of course, since the total employment in the occupation is not known, it is not usually possible to calculate unemployment rates by occupation. We do not expect to see systematic estimates of the proportions in each occupational category who are out of work. Of course, combining the data discussed in the previous two sections with the published occupational unemployment statistics, we are now able to calculate occupational unemployment rates. And hence we can ask an obvious question; is there any connection between the changes in occupational employment that we have discussed, and rates of occupational unemployment?

What sort of relationship would we expect? The most straightforward answer should simply be that if there are no effective public policies to counteract the tendency, those occupations with the highest rates of decline in employment should have the highest rates of occupational unemployment. And similarly, as unemployment rises over time we should expect unemployment in occupations with proportionately declining employment to show faster rising rates of unemployment than those in proportionately growing occupations.

It should be immediately realized that these relationships,

though they may sound tautologous, are in fact by no means so. There is in fact no necessary reason why someone who loses a job because the demand for his or her particular occupational skills has declined should be more likely to be unemployed than any other worker. There is one situation – that in which decline in employment in some occupations is not accompanied by growth in demand for others – where the connection between occupational change and occupational unemployment is in fact tautologous. But in general, the relationship between the two is mediated by the determinants of a third variable – occupational mobility. We have seen that there have been a considerable number of new jobs created in the 'knowledge occupations'. The strength of the relationship occupational change and occupational unemployment reflects therefore the extent of occupational mobility between the growing and the declining occupations.

Figure 7.1 charts this relationship between occupational change and unemployment. On the horizontal axis is plotted the change in the numbers of employees in each occupation between the two periods, with employment in 1968–73 as 100. On the vertical axis is the unemployment rate both in the earlier and the later period. A neat pattern of change emerges, with the occupational unemployment rate in each case being higher in the later period than the earlier, and with, in both the earlier and the later period, the declining occupations having a higher unemployment rate than the growing. The changes in occupational structure described in this chapter do indeed seem to bear a direct relation to the structure and distribution of unemployment.

This completes the discussion of the accounting framework. But before we can move to a consideration of its implications for the future of employment, we must consider a further set of issues; we have so far discussed, in principle, the terminants of the pattern of *demand* for labour – how the changing priorities of household needs, and social innovation, and change in the technologies and organisation of production go together to determine the occupational structure of employment and unemployment. Now we must consider labour *supply*. And as a preliminary we must approach a knotty definitional problem that we have so far evaded: the nature of 'work'.

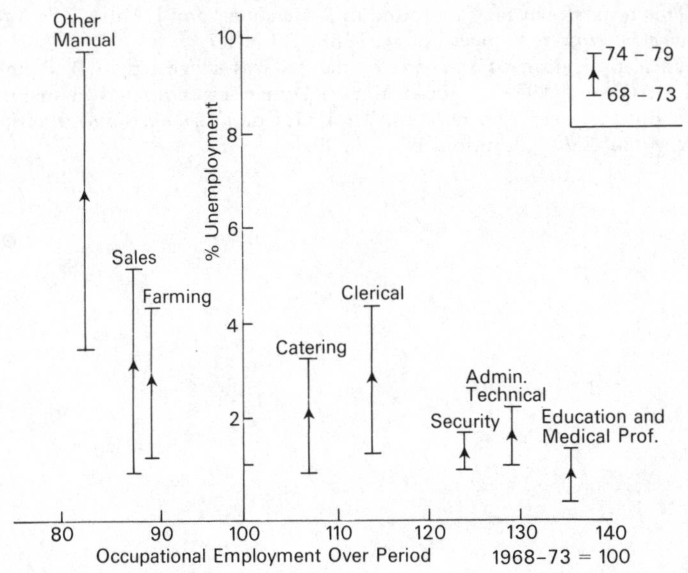

FIGURE 7.1 *Relationship between occupational unemployment rates and change in occupational employment*

NOTES

[1] Though, as we shall see in Chapter 9, the fact that carpenters are likely to be male reduces the probability of their participation in this sort of undifferentiated production activity.

[2] This two dimensional cross-classification is not very frequently used in the study of industrial structure – though tabulations in this form are available from the 1951, 1961, 1966 and 1971 Population Censuses. Guy Routh does however present data of this sort in Chapter 1 of his *Occupation and Pay in Great Britain*, Macmillan, 1981.

[3] We have no internal evidence to support the assertion that clerical workers in some sense 'service' the other categories of 'knowledge workers'. However, Crum and Gudgin, in their detailed cross-sectional analysis of the 1971 industry/occupation matrix, find that the number of administrative, professional and technical workers employed in manufacturing industries (R. E. Crum and G. Gudgin, *Nonproduction Activities in U.K. Manufacturing Industry*, Regional Policy Series 3, EEC, 1977. This associated is at least suggestive of a 'service' link between the categories.

[4] This breakdown may be found in J. Gershuny and I. Miles, *The New Service Economy*, Frances Pinter, 1983, Table 4.7.

[5] This application of shift-share analysis was suggested in D. Palmer and D. Gleave (1979) 'Labour Market Dynamics: a stochastic analysis of labour turnover by occupation', in I. G. Cullin, *Analysis and Decision in Regional Policy*, London: Pion, 1979.

Appendix to Chapter 7

There are a number of different sources of information about the distribution of employment by occupation and industry in the UK. Altogether the largest sample, and the most easily comparable long term material, is to be found in the Population Censuses for 1951, 1961, 1966, and 1971. The drawback to these data is, of course, that they are very old. To obtain more recent data, it is necessary to seek much less reliable, more aggregated material, normally only covering part of the economy.

There are five such sources. The Department of Employment publishes yearly (from 1970) statistics on the number of employees in a set of approximately 40 occupational categories – but just for the engineering industry. It also provides a series for the whole of manufacturing industry – but only giving the Administrative, Professional, Technical, and Clerical Workers, as a proportion of total employment. The Census of Production similarly provides data from which we can calculate the proportion of 'process' (i.e. manual) workers, again, just for manufacturing industry. The General Household Survey yields occupation/industry statistics, at a reasonable level of disaggregation, for the whole economy, but on a sample basis. And finally the New Earnings Survey does similarly, again on a sample basis.

After comparing the estimates from each of these sources for a number of years, it was decided that the New Earnings Survey was the best compromise between regularity between years, length of series and occupational disaggregation. The procedure by which the estimates in Chapter 7 were derived as as follows:

1. We took the estimates for total employment in each industry (i.e. 'industrial order') for each year from the Census of Employment.
2. We calculated the proportion of each industry's employment in each of the occupational categories for each year.
3. We then multiplied the total employment in each industry by the occupational proportions to estimate the number in each occupational/industrial category.
4. We finally averaged these estimates over a number of years to dispose of fluctuations due to sampling errors.

In the following pages are reproduced the estimates of absolute employment in the occupational industrial categories, as derived by this method, and also, for comparative purposes some evidence on changes in occupational/industrial proportions from the European countries, drawn from Gershuny and Miles, 1983.

TABLE A7.1 *Estimated occupation/industry distribution of employment UK ('000s of employees)*

Industries	*Occupations*									
	Admin.	Ed./Med.	Cler.	Sales	Secur.	Cat./Cl.	Farm.	Tr./Com.	Manual	All
Absolute Employment Distribution 1968–73										
Primary	416	10	460	72	8	146	3,150	412	3,796	8,468
Manufacturing	8,504	213	9,006	2,245	332	2,003	76	2,329	53,886	78,759
Utilities	1,906	10	1,646	222	54	278	38	654	12,064	16,872
All Services	12,962	14,410	26,054	13,084	2,128	19,732	1,100	8,148	17,558	115,188
Trans/Telecoms.	1,290	30	4,528	130	106	582	8	5,478	3,208	15,370
Distribution	2,978	54	4,364	10,802	56	1,300	74	1,458	5,238	26,318
Ins/Banking	1,542	22	5,786	856	70	674	84	48	478	9,556
Prof/Ed. Servs.	2,892	13,200	3,601	52	252	7,600	274	142	1,306	29,318
Misc. Serv.	1,970	654	3,146	1,240	126	7,578	214	482	4,352	19,764
Pub. Admin.	2,282	450	4,628	4	1,518	2,005	446	540	2,976	14,849
All Sectors	23,788	14,643	37,166	15,623	2,522	22,159	4,364	11,543	87,304	219,287

TABLE A7.1 *continued*

	Admin.	Ed./Med.	Cler.	Sales	Secur.	Cat./Cl.	Farm.	Tr./Com.	Manual	All
Absolute Employment Distribution 1974–8										
Primary	593	19	450	50	12	105	2,748	615	2,648	7,251
Manufacturing	9,379	320	8,594	1,704	363	2,055	75	5,190	44,677	72,413
Utilities	2,614	40	1,903	198	58	230	32	1,133	9,776	15,979
All Services	18,135	19,530	31,397	11,827	2,716	21,380	1,075	9,023	11,263	126,342
Trans/Telecoms.	1,897	53	4,486	118	135	978	5	4,229	2,709	14,609
Distribution	4,400	202	5,460	9,761	83	967	52	2,797	3,298	27,030
Ins/Banking	2,445	57	6,781	852	170	445	42	107	295	11,195
Prof/Ed. Serv.	3,370	16,983	4,841	42	92	8,338	242	225	942	35,072
Misc. Serv.	2,840	1,845	3,512	1,050	135	9,112	281	794	3,076	22,647
Pub. Admin.	3,183	390	6,317	3	2,102	1,540	453	878	943	15,788
All Sectors	30,721	19,909	42,346	13,779	3,149	23,770	3,930	15,965	68.364	221,985

TABLE A7.2 *Occupational/industrial employment as % of non-primary employment in the 1960s*

	Apt	Cler.	Sales	SCC	Trans.	Other	Total	Apt	Cler.	Sales	SCC	Trans.	Other	Total	Apt	Cler.	Sales	SCC	Trans.	Other	Total
Italy				*1961*							*1971*							*Change*			
Manufacturing	3.0	1.9	0.3	0.4	0.8	35.8	42.1	4.1	2.4	0.3	0.5	0.8	32.0	40.1	+1.1	+0.5	0	+0.1	0	−3.8	−2.0
Utilities/Con-struction	0.8	0.5	–	0.1		17.4	18.8	1.1	0.6	–	0.1		13.6	15.4	+0.3	+0.1	–	0		−3.8	−7.4
Services	11.0	4.4	2.3		21.4		39.1	13.9	6.7	3.4		20.3		44.3	+2.9	+2.3	+1.1		−1.1		+5.2
Total	14.8	6.8	2.6		75.8		100.0	19.1	9.7	3.7		67.3		100.0	+4.3	+2.9	+1.1		−8.5		
																+6.2					
Ireland				*1961*							*1971*							*Change*			
Manufacturing	1.5	2.3	0.6	0.4	2.8	19.5	27.2	2.1	2.6	0.6	0.4	2.8	19.6	28.0	+0.6	+0.3	+0.1	0	0	+0.1	+0.4
Utilities/Con-struction	0.3	0.5	–	0.1	0.3	9.4	10.7	0.5	0.7	–	0.1	0.4	11.2	12.9	+0.2	+0.2	–	0	+0.1	+1.8	+2.2
Services	21.3		14.7	11.9	7.7	6.5	62.2	23.6		12.5	9.4	7.0	6.5	59.0	+2.3		−2.2	−2.5	−0.7	0	−3.2
Total	26.0		15.4	12.4	10.8	35.4	100.0	29.5		13.2	9.9	10.1	37.3	100.0	+3.5		−2.2	−2.5	−0.7	+1.9	
																			−1.3		
UK (GB)				*1961*							*1971*							*Change*			
Manufacturing	3.6	4.3	0.8	1.1	3.2	26.2	39.2	4.5	4.2	0.8	1.0	2.9	23.1	36.5	+0.9	+0.1	−0.1	−0.1	−0.3	−3.1	−2.7
Utilities/Con-struction	0.7	0.7	0.1	0.1	0.4	7.3	9.3	1.0	0.8	0.1	0.1	0.4	6.7	9.1	+0.3	+0.1	0	0	0	−0.6	−0.1
Services	8.0	9.1	9.6	10.1	6.9	7.7	51.4	10.6	10.4	8.9	11.6	6.1	6.7	54.4	+2.6	+1.4	−0.6	+1.6	−0.8	−0.5	+3.0
															(	+4.0	)	(	−0.3	)	
Total	12.3	14.1	10.5	11.3	10.5	41.2	100.0	16.1	15.5	9.8	12.7	9.4	36.5	100.0	+3.7	+1.4	−0.7	+1.5	−1.1	−4.1	
															(	+5.1	)	(	−3.7	)	

TABLE A7.3 *Occupational/industrial employment as % of non-primary employment in the 1970s*

Industries	*Occupations* Apt	Cler.	Sales	SCC	Trans.	Other	Total	Apt	Cler.	Sales	SCC	Trans.	Other	Total	Apt	Cler.	Sales	SCC	Trans.	Other	Total
France				*1968*							*1978*							*Change*			
Manufacturing	13.7	2.1	0.4	1.8	1.0	36.7	55.7	13.9	1.7	0.3	1.4	1.2	30.1	48.6	+0.2	−0.4	−0.1	−0.4	+0.2	−6.6	−7.1
Utilities/Construction	2.7	0.4	–	0.2	1.0	11.4	15.7	3.2	0.4	0.1	0.2	0.9	7.2	11.8	+0.5	0	+0.1	0	−0.1	−4.2	−3.9
Services	11.0	2.8	1.8	4.8	1.9	6.2	28.5	16.9	3.3	2.2	6.5	3.2	7.4	39.5	+5.9	+0.5	+0.4	+1.7	+1.3	+1.2	+11.1
																			+4.2		
Total	27.4	5.3	2.2	6.8	3.9	54.3	(100)	34.1	5.3	2.7	8.1	5.2	44.7	(100)	+6.7	0	+0.5	+1.3	+1.3	−9.6	
																		(	−7.0	)	
Ireland				*1971*							*1977*							*Change*			
Manufacturing	4.1	2.4	0.3	0.5	0.8	32.0	40.1	5.8	2.4	0.5	0.5	0.8	28.6.	38.6	+1.7	0	+0.2	0	0	−3.4	−1.5
Utilities/Construction	1.1	0.6	–	0.1	0.3	13.3	15.4	0.9	0.4	–	0.1	0.3	11.1	12.8	−0.2	−0.2	0	0	0	−2.2	−3.4
Services	13.9	6.8	3.4		20.3		44.3	17.7	7.1	2.9		20.9		48.6	+3.8	+0.4	−0.5		+0.6		+4.3
Total	19.1	9.8	3.7		67.3		(100)		9.9	3.4		62.3		(100)	+5.3	+0.2	−0.3		−5.0		
UK				*1968–73*							*1974–79*							*Change*			
Manufacturing	4.1	4.3	1.1	1.1	1.1	25.6	37.3	4.5	4.0	0.8	1.1	2.4	20.8	33.6	+0.4	−0.3	−0.3	0	+1.3	−4.8	−3.7
Utilities/Construction	0.9	0.7	0.1	0.1	0.3	5.7	7.8	1.2	0.9	0.1	0.1	0.5	4.6	7.4	+0.3	+0.2	0	0	+0.2	−1.3	−0.4
Services	13.0	12.4	6.2	10.4	3.9	8.9	54.8	17.5	14.6	5.5	11.2	4.2	5.8	58.8	+4.5	+2.2	−0.7	+0.8	−0.3	−3.1	+4.0
																			−2.1		
Total	18.0	17.4	7.4	11.6	5.3	40.2	(100)	23.2	19.5	6.4	12.4	7.1	31.2	(100)	+5.2	+2.1	−1.0	+0.8	−1.8	−9.0	
																		(	−6.4	)	

TABLE A7.4 *Administrative, professional, technical/clerical ratios*

	Administrative, Professional and Technical Workers per clerical worker				Estimated Change in Clerical 'Productivity' (later/earlier)	
	1960s		1970s		1960s	1970s
France						
Manufacturing	–	–	7.4	8.4	–	+1.1
Utilities/Construction	–	–	6.5	8.7	–	+1.3
Services	–	–	3.9	5.1	–	1.3
of which: Distribution, Transport						
Finance	–	–	4.2	6.0	–	1.4
Miscellaneous Services	–	–	3.2	3.2	–	1.0
Public Administration	–	–	4.0	6.0	–	–
All Industries	–	–	5.2	6.4	–	1.2
Ireland	1961	1971				
Manufacturing	.66	.81	–	–	1.2	–
Utilities/Construction	.68	.80	–	–	1.2	–
Services	–	–	–	–	–	–
All Industries	–	–	–	–	–	–

Italy	1961	1971	1971	1977		
Manufacturing	1.6	1.7	1.7	2.4	1.1	1.4
Utilities/Construction	1.6	1.7	1.7	2.0	1.1	1.2
Services	2.5	—	2.1	2.5	.8	1.2
of which: Distribution, Transport						
Finance	—	—	1.0	1.1	—	1.1
Other Marketed Services	—	—	4.1	6.1	—	1.5
Other Non Marketed Services	—	—	2.5	2.7	—	1.1
All Industries	2.2	2.0	2.0	2.4	0.9	1.2
UK						
Manufacturing	0.	1.12	.97	1.13	1.2	1.2
Utilities/Construction	1.3	1.3	1.15	1.4	0	1.4
Services	1.05	1.20	1.05	1.20	1.1	1.1
of which: Producer Services	—	—	0.40	0.54	—	1.3
Public Consumer Services	—	—	4.5	4.2	—	0.9
Private Consumer Services	—	—	0.84	1.3	—	1.5
Public Administration	—	—	0.59	.56	—	0.9
All Industries	1.03	1.10	1.03	1.17	1.2	1.1

TABLE A7.5 *Occupational vs. industrial shifts*

	White Collar	Other Service	Other Non-Service	White Collar	Other Service	Other Non-Service
		1960s			*1970s*	
France						
1. Earlier Occupational Distribution	–	–	–	32.7	12.9	54.3
2. Industrial Shift	–	–	–	+2.5	+2.6	−5.1
3. Occupational Shift	–	–	–	+4.2	+0.5	−4.5
4. Later Occupational Distribution	–	–	–	39.4	16.0	44.7
Ireland						
1. Earlier Occupational Distribution	26.0	38.6	35.4	–	–	–
2. Industrial Shift	−0.8	−1.6	+2.5	–	–	–
3. Occupational Shift	+4.3	−3.8	−0.5	–	–	–
4. Later Occupational Distribution	29.5	33.2	37.3	–	–	–
Italy						
1. Earlier Occupational Distribution	21.6	3.9*	74.3*	28.9	5.1	65.0
2. Industrial Shift	+1.6	+0.2	−1.7	+1.5	+0.2	−1.6
3. Occupational Shift	+5.7	+1.0	−6.7	+3.9	−0.2	−3.7
4. Later Occupational Distribution	28.9	5.1	65.9	34.3	5.1	60.6
UK						
1. Earlier Occupational Distribution	26.4	32.3	41.2	35.4	24.3	40.2
2. Industrial Shift	+0.5	+1.2	−1.6	+1.0	+1.2	−2.0
3. Occupational Shift	+4.6	−1.5	−3.1	+6.3	+0.5	−6.9
4. Later Occupational Shift	31.5	32.0	36.5	42.7	26.0	31.3

CHAPTER 8

The Supply of Labour 1: A Diminishing Marginal Utility of Income

8.1 Who Works?

ANTHROPOLOGISTS observe that in some pre-industrial societies, 'work' is not a concept that has any very great discriminatory power. Activities which we, from an industrialized society, would describe as work, are not in any way distinct from other activities which we would not so describe. All activities, in such societies, are so bound up in a complex pattern of ritual and sociability, that each is seen as being undertaken for its own intrinsic purpose. This is not to say that there are no instrumental purposes, that no present activities help to achieve more distant aims – but rather that no one engages in what we might see as work activities *because* they achieve such extrinsic purposes. Each daily task has a bundle of meanings for the various parts of an individual's social and economic life, which go, for the most part, unanalysed.

Technical change and the division of labour break into this complex web of latent functions. As work tasks are reorganized into new specializations, so individuals find themselves involved in new tasks which are explicitly justified in terms of their instrumental significance. Instead of doing things because they are the right activity for that individual at that time, and his friends or brothers or uncles are doing the same thing in his company, he works at the newly specialized task *in exchange for* some specified future benefit.

Of course, any particular set of specializations can become ritualized. Complex networks of originally instrumental tasks *can* become so habitual that the original rationale is submerged within a stable, and culturally transmitted, pattern of behaviour. So, obviously, the crucial issue is not the division of labour itself, but *change*. Technical innovation leads to new task specializations, which in turn call for new instrumental evaluations by those who are to do the tasks.

So here we have one reason that 'work' emerges as a distinct entity in industrial societies. A continuous process of change in the techniques of production prevents the incorporation of tasks into an organic 'seamless web' of activities undertaken for unspecifiable reasons, and maintains them among those activities undertaken for a single, specifiable, instrumental purpose.

And there is a second reason, associated not with change, but with the division of labour itself. Specialization may produce *nasty* specialities – dirty, repetitive, meaninglessly trivial – such that no habituation could persuade the specialized worker that he or she engages in them for any reason other than their instrumentality.

The first of these two explanations clearly applies to those work activities for which we are paid; but what should we say about unpaid work? Why, for example, do people consider domestic activities to be work? The second explanation seems to apply; for a number of reasons, people *dislike* housework. It is normally isolated. The tasks are sometimes intrinsically unpleasant; and while they might be bearable if we shared them equally, when some of us are forced to specialize in them, they are not. And specialized household tasks have low ascribed status, which adds to their unpleasantness. These circumstances mean that housework is not in general done for its own sake, but for the external purpose of the well-being of the household – again an instrumental purpose.

The instrumental nature of work serves as a convenient means of definition. The Canadian economist Oli Hawrylyshyn defines work ('economic activity') according to what he terms the 'third party criterion': 'An economic activity of any individual is one which can be done by a third party (generally hired at a market price) without affecting the utility value to the individual.'[1] This enables us to distinguish between activities such as scrubbing floors, which we could reasonably pay someone else to do for us, and activities such as watching a football match, for which the purchase of a proxy would be, except in very special circumstances, ridiculous. The act of scrubbing the floor generates in itself principally an indirect utility, it is a means, the end being, if the person doing the scrubbing owns the floor, a clean floor (i.e. a direct utility) and if not, money with which to purchase a direct utility.

The test for work is twofold – work generates some indirect utility *and* can be purchased. This test identifies, for an industrialized society, a clear set of activities, some of which are predominantly undertaken in exchange for money payments, and others which are predominantly unpaid.

But though the criterion is a successful definitional device, it is not, sociologically speaking, very insightful. Though jobs may be undertaken in exchange for pay, some people enjoy doing them (nevertheless, it is only a very small, and probably mendacious, minority who would claim they would stay in their jobs if they were *not* paid). And, probably more important, we must add that the apparent instrumental purpose of jobs may mask other underlying 'latent' functions that the work fulfils for the worker. Marie Jahoda[2] identifies a set of such functions for paid work – including the provision of a temporal organization, frameworks for sociability, and the ascription of status, physical exercise – and suggests that it may be the loss of these that causes the frequently observed physical and psychological breakdowns among unemployed workers.

It is not usual to include unpaid work in discussions of 'the supply of labour' – but it is necessary for the argument developed here. Most of the recent increase in paid work comes from people previously engaged in unpaid work. And, it can be argued, much of the unequal distribution of paid work between men and women is a consequence of the unequal distribution of unpaid work between them. More women in the paid work-force has meant an inequitable *redistribution* in the labour supply, rather than an increase in it.

By now it is obvious that the answer to the question who works? is, essentially, everyone. This chapter will look at changes in the disposition of the work of the adult population, using conventional sources for data on the paid work-force; and Chapter 9 will use rather less conventional 'time budget' data as a basis for considering unpaid work.

8.2 Who Have Paid Jobs?

Figure 8.1 gives a long-term picture of the pattern of change in overall participation in paid work. It shows rather different developments for the sexes. The male participation rate shows a very regular decline from 1931. The 10–20 per cent not participating in the work-force represent almost entirely retired men who have previously been in the work-force; the decline in the average rate of participation reflects simply the increasing average age of the male population resulting from the increasing size of the over-65 age-group.

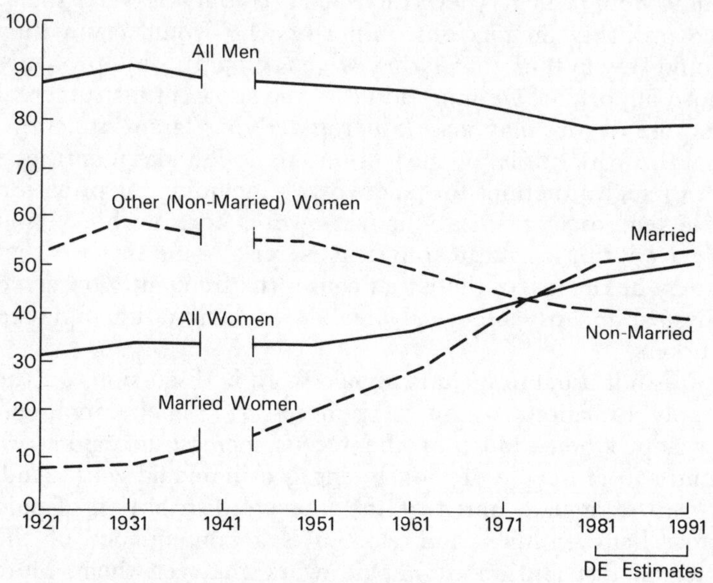

FIGURE 8.1 *Adult participation in paid work, GB, 1921–91*
Note: Includes all aged 12+ in 1921
14+ in 1931
15+ in 1951 onwards
married females 15+ in 1921
16+ in 1931 onwards
Excludes full-time students.
Source: Department of Employment *Historical Abstract of Statistics* (*1921–61*), and Department of Employment *Gazette*, Apr. 1978 (1971–91).

While almost all the men have jobs, only about half the adult women do. There has been a relatively slow but quite regular increase in the average rate of participation of women, from about 32 per cent in 1921 to about 48 per cent in 1981. But the really striking change emerges when we look separately at the statistics for married women as against others. Married women's participation rate increased from about 8 per cent in 1921 to around 51 per cent in 1981 – with about half the increase concentrated into the past two decades. Until the 1950s unmarried women made up the great bulk of the females in employment. The decline in the average participation of the non-married group over the period reflects three separate demographic changes: more women marry – and marry young – than previously; more old unmarried women live long into retirement; and more old women who have been married survive their husbands. So the curve for 'other women' represents a population whose age profile becomes very considerably skewed as we move along it. If we were to correct for this demographic change, we would probably find that, age for age, the participation rate of non-married women has risen somewhat – though of course, not at anything like the rate at which married women's participation has increased.

So in sum, we have men with a somewhat declining average participation rate, because of demographic change, and women with a very fast increasing rate of participation, in spite of demographic changes which on their own would have reduced their participation rate faster than that of men. This pattern is a pretty general one for the developed world as a whole. As Table 8.1 sets out, 12 of the 20 countries covered showed the same pattern as the UK from the mid-1960s to the mid-1970s, while none showed the contrary pattern of male employment increasing and female decreasing.

This presentation obscures another major difference between male and female patterns of participation – the extent of part-time work. Only a small part of the male workforce works part-time – 3.6 per cent in 1971 growing to 4.3 per cent in 1977. For women, by contrast, part-time workers are a very significant part of the workforce – 30 per cent in 1971, and 36 per cent in 1977. As Figure 8.2 suggests, part-time work tends to increase in step with the increase in the overall participation rate. In fact, over the period covered by this figure, female employment in the UK increased by 1.03 m., of which .86 m. were part-time workers; so, in net terms, 83 per cent of the increase of female participants over the period came from part-time workers. If we may turn for a moment from longitudinal to cross-sectional evidence, we can see from Figure 8.3 that this fits quite well into a European pattern –

TABLE 8.1 *Change in participation in paid work OECD 1966–77.*[a]

(Data not available for Iceland, Luxembourg, Netherlands, or Switzerland).

$$\text{Participation Rate} = \frac{\text{Total Labour Force}}{\text{Total Population}} \times 100$$

	Change in Participation						
	All	Men	Women	M = + F = +	M = – F = +	M = – F = –	M = + F = –
Canada	+	+	+	√			
USA	+	+	+	√			
Japan	–	–	–			√	
Australia	+	–	+		√		
New Zealand	+	–	+		√		
Austria (1969–77)	–	–	–			√	
Belgium	+	–	+		√		
Denmark (1967–77)	+	–	+		√		
Finland	–	–	+		√		
France (1968–75)	+	–	+		√		
Germany	–	–	–			√	
Greece (1961–71)	–	–	–			√	
Ireland (1966–73)	–	–	–			√	
Italy	–	–	+		√		
Norway	+	–	+		√		
Portugal	+	–	+		√		
Spain (1970–6)	–	–	+		√		
Sweden	+	–	+		√		
Turkey (1960–70)	–	–	–			√	
UK	–	–	+		√		
				2	12	6	0

[a]OECD Manpower Statistics

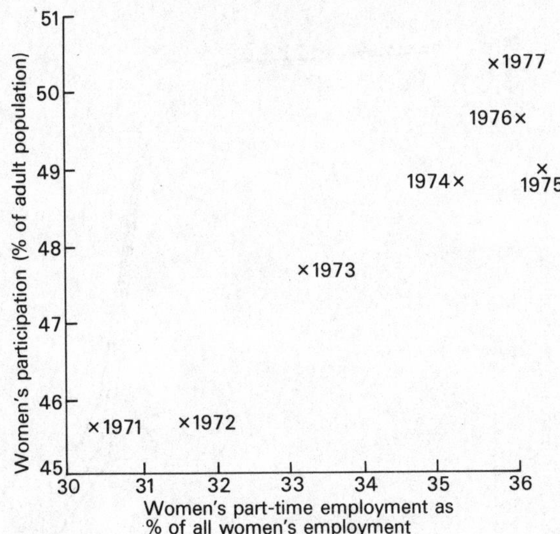

FIGURE 8.2 *Relationship between women's participation in paid work and part-time paid work, UK.*
Source: *Census of Employment*, and Department of Employment, *Gazette.*

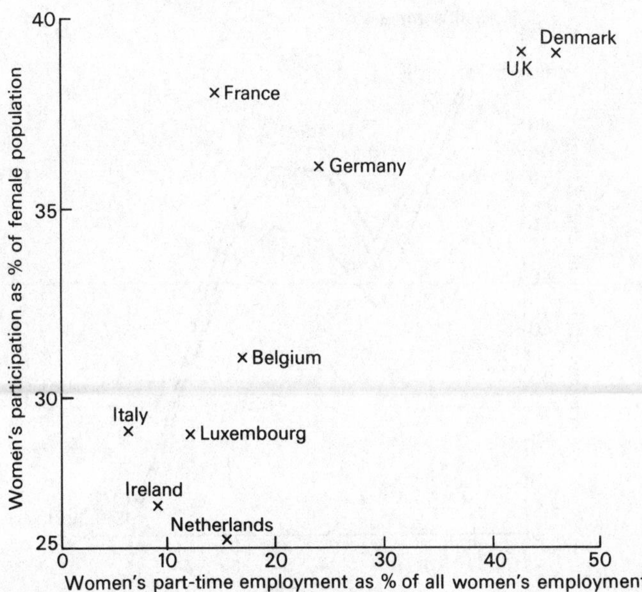

FIGURE 8.3 *Relationship between women's participation in paid work and part-time paid work*
Source: EEC Labour Force Sample Survey, 1977

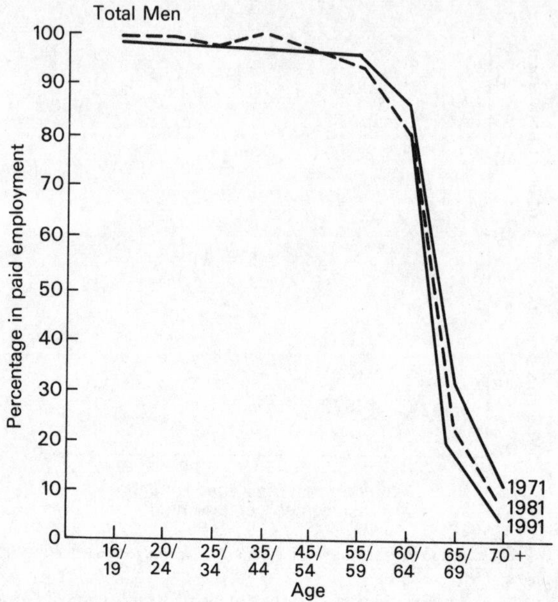

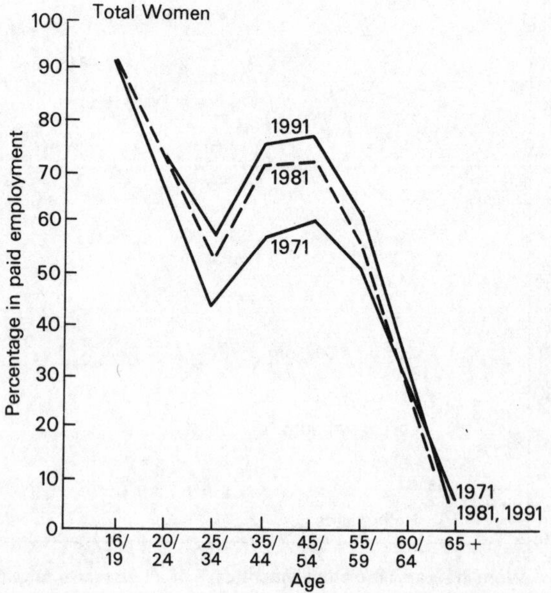

FIGURE 8.4 *Participation in paid work by age, 1971, 1981, 1991*
Note: Full-time students classed as economically *active*.
Source: Department of Employment, *Gazette*, Apr. 1978.

the higher the female participation rate, the higher the proportion of part-time workers in the female workforce.

We begin to get a picture of what underlies this difference between men's and women's participation rates when we look at their participation rates broken down by age (Figure 8.4). Essentially, *all* men participate in the paid work-force (for these purposes we may include full-time students among the employed), until old age or disability get to them; the pattern has hardly changed over the past decade, nor is it expected to change very considerably over the next decade. Women, however, show a much more varied pattern, with a participation rate of above 90 per cent for the 16–19 age group, falling sharply in the decade between 25 and 35, then rising again until their mid-50s. And the participation rate among the 35–55 age group has risen markedly over the past decades, and is expected to continue to rise. The pattern becomes clearer in Figure 8.5, where we disaggregate women into the married and non-married categories. Taking non-married women first, we find an initial relatively small fall in the participation rate, until the mid- to late 20s, presumably reflecting some women leaving work to care for aged relatives, or to enter functional equivalents to marriage which do not appear in the official statistics. Then, from the late 20s to the mid-50s, the participation rate stays approximately constant at around 75–80 per cent, after which it declines with retirement.

When we compare the two graphs in Figure 8.5, it becomes immediately apparent that the transition from the non-married to the married state for young women virtually halves their probability of participation in paid work. What is it about marriage that makes women leave their paid jobs? Part of the answer emerges from the next figure (though the full argument will not emerge until we consider the distribution of domestic work in Chapter 9. Figure 8.6 shows the relationship between responsibilities for childcare and participation, part- and full-time, in paid work.

The message carried in Figure 8.6 may be summarized as follows:

1. It shows that the older the family, the more probable it is that the mother will have a paid job, and, for those in employment, the more probable it is that the job is a full-time

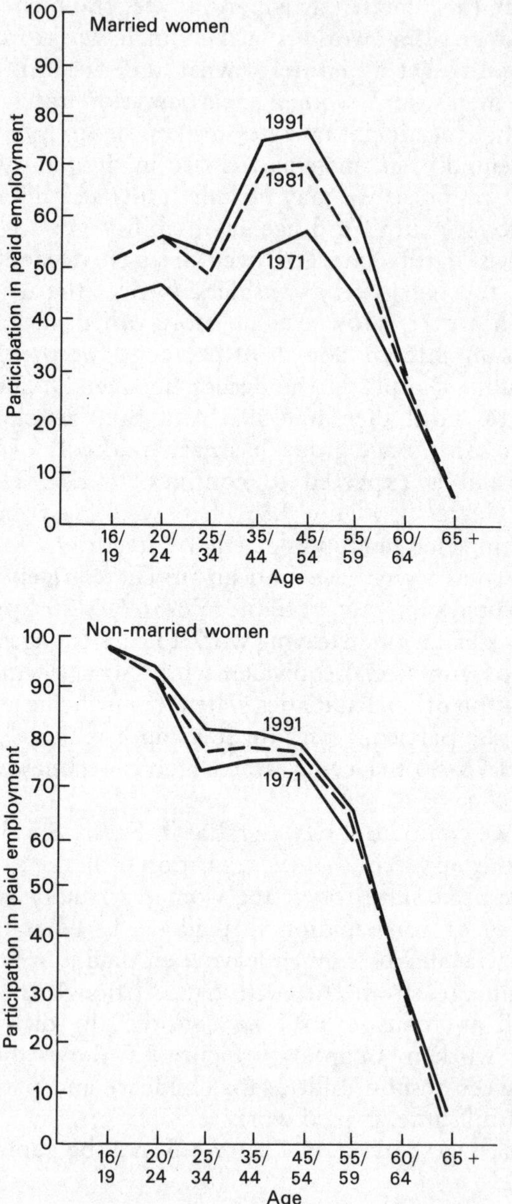

FIGURE 8.5 *Participation in paid work by age, 1971, 1981, 1991*

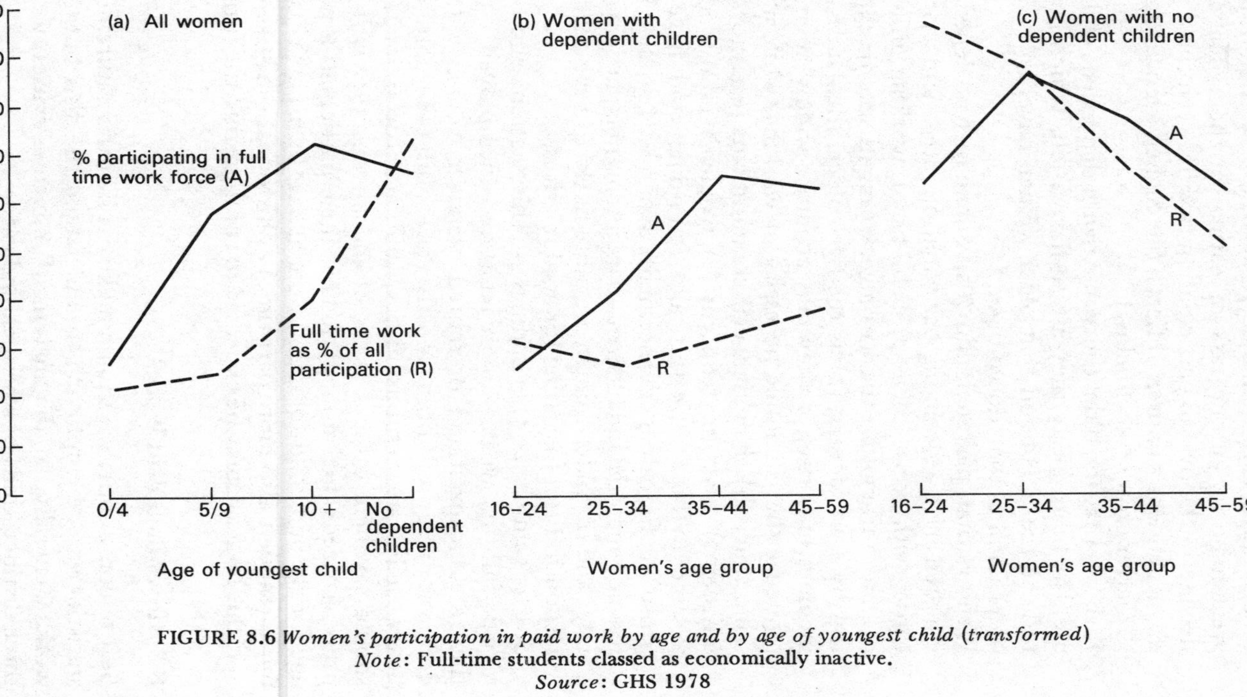

FIGURE 8.6 *Women's participation in paid work by age and by age of youngest child (transformed)*

Note: Full-time students classed as economically inactive.

Source: GHS 1978

one. This presumably reflects (a) that younger children require more care than older, and (b) that women bear the great majority of the task of caring for them. The slight decline in the overall participation rate between those with no dependent children reflects the higher average age of the latter group (see 3 below).

2. It shows that the older the women with dependent children are, the more likely they are to have a job. Which merely reflects the likelihood that older women have, on average, older families than do younger.
3. It shows that, past their mid-30s, women with no dependent children have a declining participation rate – and that, for those with jobs, their probability of working full-time declines throughout the age range. We might seek to explain this by four factors; (a) as they get older, their husbands' incomes may rise, or they may acquire a widow's pension, reducing the economic necessity of working, while at the same time, (b) their financial commitments (e.g. mortgage payments) may be reduced; (c) women's jobs tend on average to be less intrinsically rewarding, and have less prospects for career advancement, than men's, so they have less non-material inducement to maintain their participation rate; and (d) older women are more likely to have acquired responsibilities for caring for aged relatives.

So, to summarize the findings of this section: almost all men do full-time work – or are registered unemployed – until they are incapacitated or retired; younger women without domestic responsibilities mostly have full-time jobs, but heavy child-care responsibilities, and age (or perhaps we should say here other sources of income together with less pressing needs for money) reduce their participation; and, age-for-age and child-for-child, women have been participating more (though also more part-time) as historical time passes – and this increase may be expected to continue into the future.[3]

8.3 How Much Paid Work?

Before continuing, we should make explicit the nature of the statistics we are employing to understand the distribution of work. Generally, in the analysis of human activities, we use three distinct kinds of statistical indicator:

– statistics about rates of participation in the particular activity ('P-statistics')
– statistics about the average time spent in the activity by *each participant* ('T′-statistics')
– and statistics about the *average* time spent in doing the activity by a population or sub-population, *irrespective of whether or not particular individuals actually participate in activity*. ('T′-statistics')

Obviously, the product of the first two of these must be equal to the third of them, i.e.:

$$T = PT'$$

In the previous section we considered P-statistics for paid work; in the following paragraphs we shall consider T′-statistics, also for paid work; in Chapter 9 we shall consider the statistics for unpaid work. Figure 8.7 gives a picture of change in the average hours of work of full-time manual workers between 1948 and 1979. It shows a regular (if stepwise) decline in hours of work since 1955.

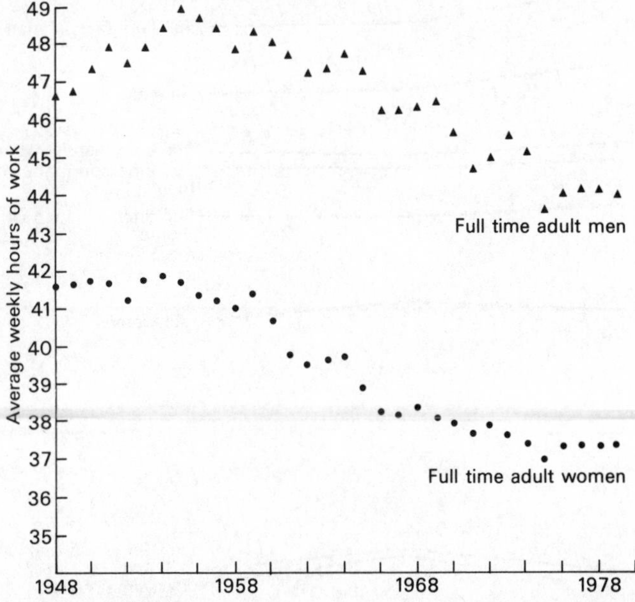

FIGURE 8.7 *Actual hours of work, full-time manual workers, 1948–79.*
Note: Hours actually worked but excluding firms with short-time working.
Source: Department of Employment, *Abstract of Historical Statistics*, and Department of Employment, *Gazette*.

The decline is by no means a trivial one. Average male manual-working hours have declined by something of the order of five hours from their post-war peak (which is also quite representative of pre-war work-patterns). Women's full-time work was reduced, over the same period, by rather more than four hours per week.

We can get a more general picture for the UK in the 1970s across the full range of occupations, from the New Earnings Survey (started in 1968, to replace the Department of Employment's long standing 'voluntary survey of earnings and hours' from which the Figure 8.7 comes). Figure 8.8 shows a rather higher absolute number of hours for manual men and women than that indicated by the previous figure. But the scale of the declines in working time – of rather more than an hour per week for men, and three-quarters of an hour for women, during the 1970s – are of the expected magnitude.

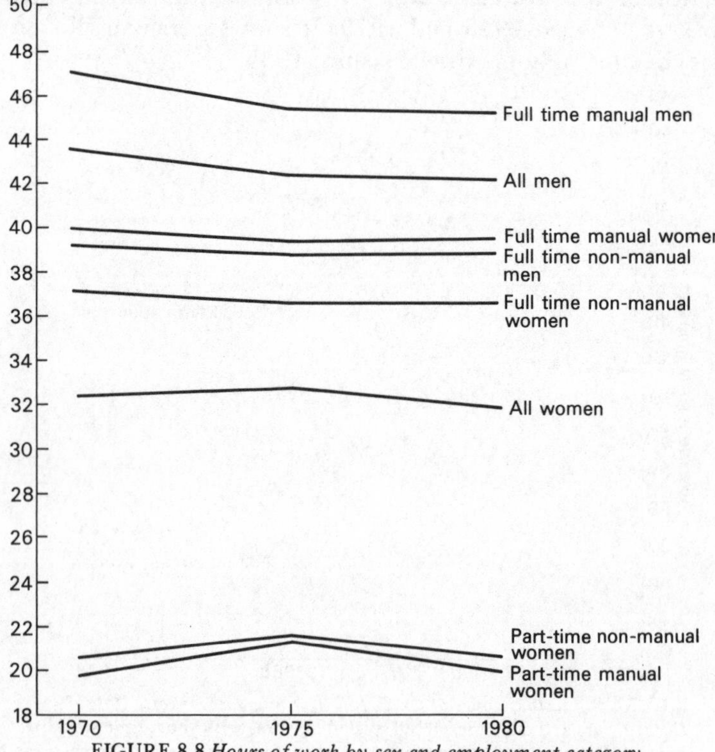

FIGURE 8.8 *Hours of work by sex and employment category*
Source: NES 1970, Table 103; NES 1975, Table 26; NES 1980, Table 27.

We see that most of the categories of paid workers have reduced their hours; the only marked exceptions to the general decline being part-time women who showed, in both the manual and non-manual categories, a very substantial increase in working hours between 1970 and 1975. This presumably reflects the very considerable growth in the level of employment (the P-statistic) during these years, which must have been associated with a widening in the range of jobs offering a part-time option. (It should be added that the definition of part-time work also changed over this period.)

But let us return for a moment to the data in Figure 8.7. The decline in working time shown here is regular, but in no way smooth. Hours of work decline in a series of four-to-five yearly jumps, which are separated by periods during which hours of work stay approximately constant or even rise a little, giving a ratchet-like decline in hours, with the same periodicity as that of the short-term business cycle. We can illustrate this by comparing the yearly change in hours of male manual workers with the yearly change in their real wages (Figure 8.9). We can see that there does seem to be, at least during the 1960s, a similar sort of cyclicity in the two series. We cannot tell anything very definite about the pattern of causation here, but we do have two alternative explanations:

1. The fast rate of decline of hours are associated with *simultaneous declines* in the rate of increase of real incomes (i.e. with the downturn of the business cycle employers are willing to cede, and employees are forced reluctantly to accept, shorter working hours but no increased rates of pay).
2. The fast rates of decrease of hours are *lagged* effects of previous fast rates of *increase* in income (i.e. workers who have previously received increases in their real income, are now seeking increased leisure in which to enjoy it).

There will certainly be elements of truth in both sorts of explanation; but a moment's thought will demonstrate that the second must be the dominant process.

Consider: if the first held, but the second did *not*, then, with the return to high output growth in the upswing of the next business cycle, hours of work should return to their previous level – the 'yearly change in hours of work' in Figure 8.9 should

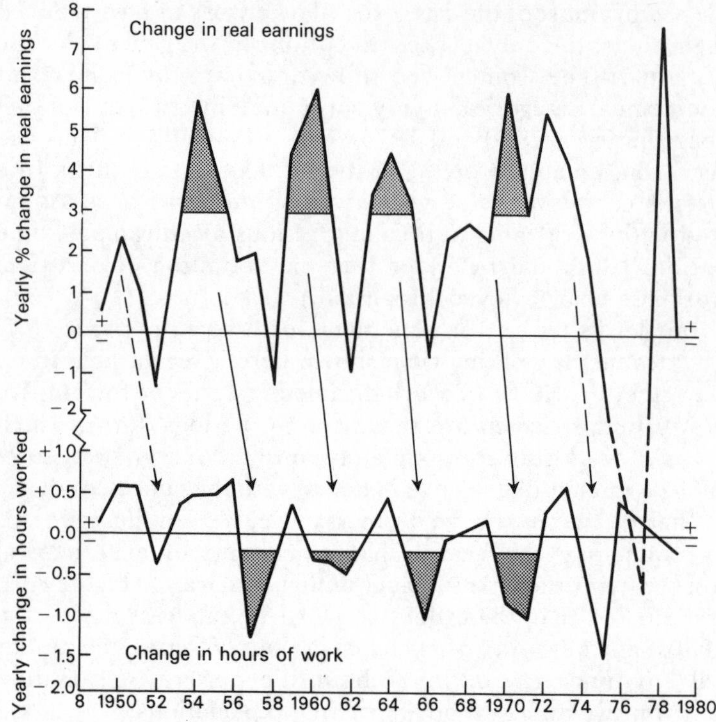

FIGURE 8.9 *Change of real earnings and hours, male manual workers*
Source: **Department of Employment,** ***Abstract of Historical Statistics*****, and Department of Employment,** ***Gazette*****.**

oscillate *symmetrically* about the origin. It does not do so; the values below zero comfortably outweigh those above it. Hence the *downward* ratchet in Figure 8. . So we can only conclude in favour of the second hypothesis, that people are choosing leisure rather than taking all the benefits of economic growth in the form of income. This does not, of course, invalidate the first hypothesis – but it does become somewhat downgraded into a *facilitator* of the second. The *mechanism* by which shorter hours are achieved is the willingness of employers to concede shorter hours rather than higher pay during a cyclical downswing.

We should notice also that this neat model breaks down as we progress through the 1970s. As economic growth effectively ceases, so hours of work initially fall (faster even than in the

1960s, encouraged by employers' labour-hoarding and government job-protection schemes) but then maintain themselves or even rise slightly as workers attempt to protect themselves against loss of real income as the economy falls into a longer-term recession. Nevertheless, hours of paid work show no sign of returning even to the level of the 1960s. We might also note at this point that the 'ratchet' process may provide some sort of explanation for the frequently expressed assertion that hours of work are not, in general, declining. For the last twenty-five years or so, four out of every six years show relative constancy or even small increases in hours – so looking at short time periods will tend to disguise the process of decline.

These arguments bring us up against what is now considered a rather old-fashioned economic concept – the diminishing marginal utility of income (DMUI). It is true that the concept does not help us in any way to *explain* what happens to labour supply, and modern economists use the analytically more elegant notions of 'income' and 'substitution' effects (which respectively lead to decreases and increases in the amount of labour supplied as a consequence of an increase in the wage rate) in their models. But nevertheless, DMUI does seem to be a useful *descriptive* notion in the context of the data we are considering. So let us be precise; DMUI is used to denote a situation in which, given an increase in the productive potential of a society, people are willing to take only a part of this increased potential in the form of an increase in real income, and wish to take the remaining part in the form of increased leisure time. This is, of course, distinct from a *negative* marginal utility of income, which would imply a wish for an absolute reduction of real income in exchange for more leisure. We find no evidence for any such wish – but the evidence does point to DMUI; over the long-term, rising income is associated with shorter hours of work.

There is still one objection that might be raised to this argument. The DMUI conclusion rests on the imbalance between the large decreases in the hours of work in the downturn of the business cycle, as compared with the relatively small increases associated with the upturn. Sceptics might argue that this relates, not to labour supply factors, but to labour demand; it may be that British industry was simply not able to take sufficient advantage of upturns in the business cycle to use all the labour that might have been available. In other words, the phenomenon may reflect the specific British failure to manage successful growth of output during the 1960s. We can only counter this objection by reference to other, more successful countries. We find that declining hours of paid work is a general phenomenon throughout the richer countries; take Japan, for example, the most successful economy in the 1960s. Average monthly hours of work declined from 203 in 1960, to 193 in 1965, to 185 in 1970, and the decline continued through the 1970s. If the reduction in working hours shows up in successful economies *as well* as in the UK, we must conclude that it reflects the wishes of the workers rather than the inefficiency of the management.

If we now move from the consideration of time-series 'longitudinal' data, to cross-sectional data, we can perhaps throw some more light on this phenomenon. Figure 8.10 shows, for three different dates, the average amount of work done by adult men paid at various different hourly wage rates. For each of the three years, we initially see a smooth increase to something around 45p per hour (in 1970 prices, equivalent, that is, to about £1.50 per hour in current prices), and thereafter a smooth decline in hours worked.

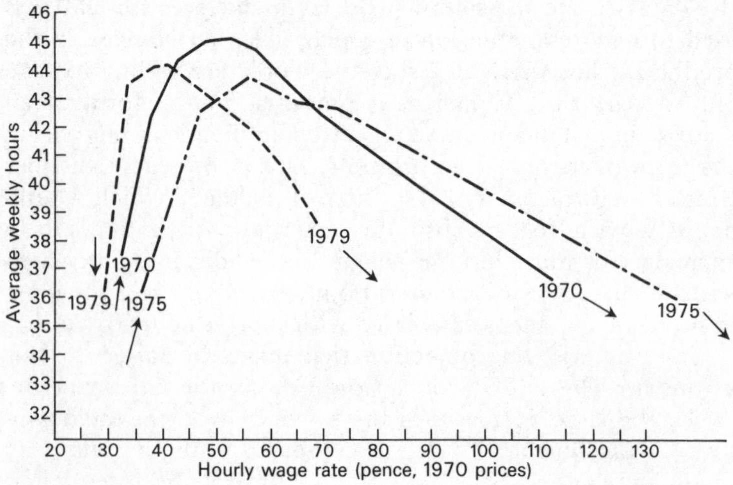

FIGURE 8.10 *Average weekly paid hours of work by hourly wage-rate (hourly-paid adult men)*
Source: NES 1970, Table 120; NES 1975, Table 161; NES 1979, Table 161.

There are two reasons for the differences between these curves. The first is that inflation through the 1970s was very high, and the Retail Price Index, which we use here as the price deflator, is not reliable for this purpose; small differences in the curves may not, in fact, be significant. If, for example, we had used a deflation factor of 2.9 rather than 3.1 for the 1979 data, the maxima of the 1970 and 1979 curves would match almost exactly. Second, the information comes from 'grouped data' published by the *New Earnings Survey*; this publication groups the data in a different way each year. For obvious reasons, we have not plotted the final group (i.e. that group 'earning *more than* X pence per hour'). In 1979 this group contained 11 per cent of the working population, whereas in 1975 the equivalent group only contained 3 per cent of employed males. So the 'tails' of the distribution are not comparable.

We can, therefore, to some extent disregard the differences between the curves for the three years; what catches our eye is the similarity between them. They really do seem to exhibit a behavioural regularity. Now, it could be argued that the common shape of this curve reflects a labour *demand* factor. But to do so, we would have to suppose that employers are more likely to give overtime, and high basic hours, to mid-priced workers rather than to either high or low wage workers. This is possible but, on the face of it, rather unlikely. It would seem more reasonable to say that the shape reflects conditions of labour supply, that workers decide on their desired hours of work on the basis of the wage rate. And the general shape of the curve, bending continuously to the right as it does, is the one that would follow from a DMUI.

The equivalent women's curve Figure 8.11 suggests rather different conditions of supply. The curve does not turn down until it reaches 60p per hour – and even then it does not do so in any very determined manner. We might note that the right-hand tail of the distribution is shifting upwards, which may reflect a demand factor – employers being more willing to give overtime to worse-paid women than to better-paid men. In general, the curve is flatter and more irregular than that for men. This may in part reflect the smaller sample of women in the *New Earnings Survey*. But more generally this is probably because there are *other* determinants of women's

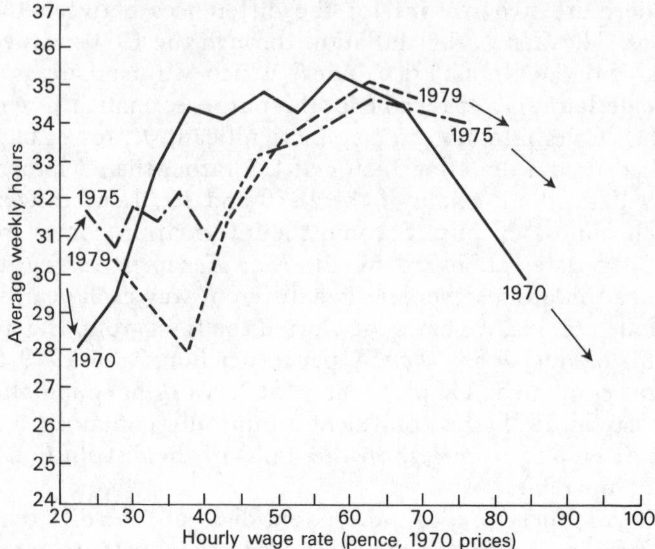

FIGURE 8.11 *Average weekly paid hours of work by hourly wage-rate (hourly-paid adult women)*
Source: NES 1970, Table 121; NES 1975, Table 162; NES 1979, Table 161.

labour supply than the wage rate – which are (as we argued before in relation to T-statistics) the range of non-paid work responsibilities that women bear.

But before we consider unpaid work, let us take the argument concerning DMUI a little further. Working from Figure 8.9 we can calculate the (very) approximate weekly income of each wage group by multiplying the mid-point of the wage-range by the average working hours of that group. The results of this process are exhibited in Figure 8.12. We must note something about the distribution that we would get if we were simply to plot income against hours of work. A high income can be earned either by earning a high hourly wage for relatively short hours, or a low wage for very long hours. Since, for any given wage rate, there is a very wide dispersion of hours actually worked around the average value that we have been discussing in the previous few paragraphs, a direct plot of income against hours would show no very clear relationship since each income level would contain a multitude of different wage rates.

In other words, in Figure 8.12 we are still really looking at

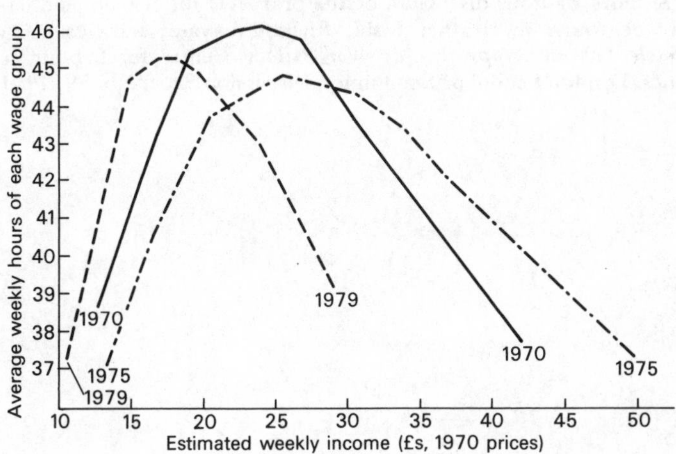

FIGURE 8.12 *Average weekly paid hours of work by estimated average weekly income of the wage groups*
Source: NES 1970, Table 120; NES 1975, Table 161; NES 1979, Table 161.

average hours of work for different wage-rate groups. But the point to stress is that this form of grouped data is the *correct* place to find evidence of DMUI. If we were to look directly at the plot of income against hours (and particularly if we were to do so for a particular occupational stratum) we would probably find on average an *increase* in hours of work with total income – simply because you need to work long hours (particularly at a given wage rate) to earn a high income. The direct comparison of total incomes with hours of work is simply an *incorrect* basis for exploring the Marginal Utility of Income.

So, two quite clear conclusions emerge from this chapter. First, it appears that in principle *all* adults wish to (or feel obliged to) participate in paid work. Second, the *amount* of paid work they wish to do tends to decline as income increases.

NOTES

1 Oli Hawrylyshyn, *Estimating the Value of Household Work in Canada, 1971*, Statistics Canada, 1978, p. 17.

2 Marie Jahoda, *Employment and Unemployment: A Social Psychological Analysis*, CUP, 1982.

[3] A more rigorous discussion of the prospects for female paid labour may be found in Heather Joshi, Richard Layard, and Susan Owen, *Female Labour Supply in Post-War Britain*, Centre for Labour Economics, London School of Economics, Discussion Paper No. 79, 1981.

CHAPTER 9

The Supply of Labour 2: The Distribution of Unpaid Work

9.1 How Much Domestic Work?

In the previous chapter, the empirical discussion of paid work was conducted entirely in terms of participation rates and participation time (P- and T′-) statistics. The arguments in this section involve inherently more complex phenomena – such as the effect of participating in one sort of activity on time spent in another activity, and the effect of a wife's participation in an activity on her husband's behaviour – so to continue to consider labour supply in terms of P- and T′-statistics would be to make the discussion unmanageable. Accordingly, most of the arguments will concern the simpler population average time (T-) statistics, drawn from a 'time budget'[1] study carried out at the Science Policy Research Unit.

But to serve as a transition from the previous discussion, let us first consider the P- and T′-statistics for paid and unpaid work, as they appear from the time budget estimates (Figure 9.1). The participation rate estimates for paid work are rather higher than those which emerge from the official statistics; only a little higher in the case of men (88 per cent as against 86 per cent in 1961, 86 per cent as against 83 per cent in 1974/5) and very considerably higher (44 per cent as against 38 per cent in 1961, 57 per cent as against 44 per cent in 1974) for women. But these discrepancies can be very easily explained: 'participants' from the time-budget data include all those who admit in their diaries to doing even half-an-hour's paid work. Many of the very short-time participants by this definition would not be identified through official sources. And women, who show, as we have seen, particularly fast rates of increase of part-time work, would be expected to be particularly badly underrepresented for this reason – hence the larger difference, and the faster rate of growth of participation emerging from the time budget data. The T′-statistics for paid work are also a little higher than would be expected from the official data – but here we have a more directly

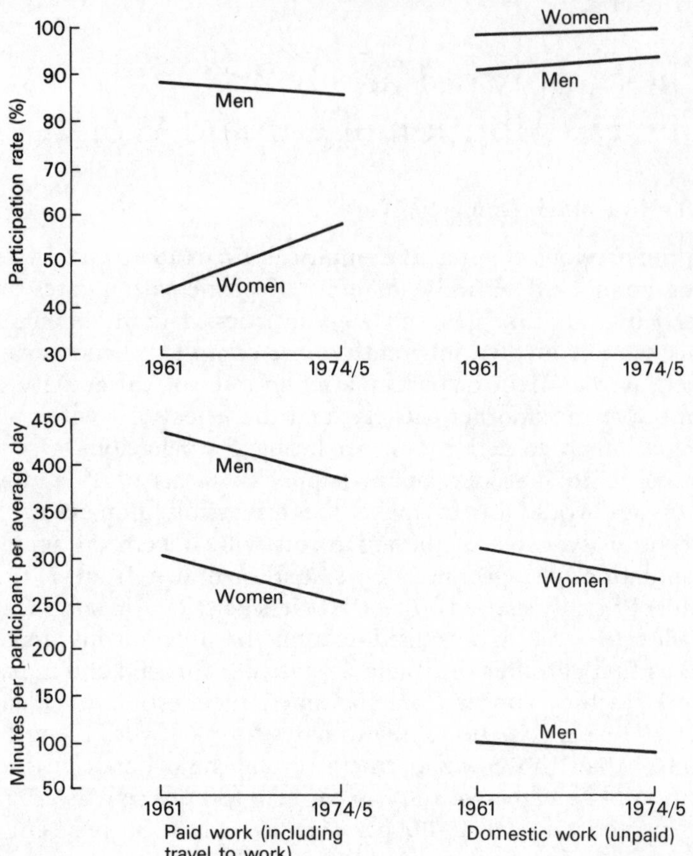

FIGURE 9.1 *P- and T'-statistics, paid and unpaid work, for men and women, 1961 and 1974.*

checkable explanation; because of shortcomings in a number of the diaries, we have to include 'travel to work' in our estimate of paid work time. When we correct for this, the time budget and the official estimates coincide almost precisely (Gershuny and Thomas, 1980, pp. 100–3).

Of course, of more interest to us are the estimates of change in unpaid work,[2] which are not available to us from any other source. Virtually all women (99 per cent in 1961, 100 per cent in 1974/5) participate in domestic work – and, by 1974/5, very nearly all men did so (94 per cent in 1974/5). The very big difference between the sexes comes from the participants' time (T′-) statistics. Men do roughly one-third as much housework as do women; and if there was any redistribution of housework over the period, it came from women doing less housework rather than men doing more.

9.2 The Domestic Labour Paradox

Before we go any further, let us consider what our previous arguments would lead us to expect about the change in unpaid work over time. Thinking for a moment rather loosely, we might say that over the last decades we have seen changes in the mode of provision for various final service functions, in the direction of more informal production – hence we would expect more informal work time. And indeed it has frequently been observed that housework increases over time in spite of the diffusion of 'labour saving' devices. Consider for example the USA data presented in Table 9.1; the distribution of housework activities has changed somewhat over the period, but it is nevertheless clear that housework time in total increased over the period for employed and non-employed women alike.

TABLE 9.1 *Comparison of time spent in household work between 1952 and 1967/8 (USA)*

	(Minutes per average day)			
	Employed Women		Housewives	
	1952	1967/8	1952	1967/8
All food activities	114	96	156	138
Care of House	48	72	96	96
Care of Clothes	48	54	96	78
Care of Family Members	18	48	66	108
Marketing and Recording	18	48	30	60
All Household Work	246	318	444	480

It seems quite likely that we can generalize from this result. An analysis of the data on housework from the multinational time budget study carried out under the auspices of UNESCO during the mid-1960s leads to the following conclusions:[3]

There is little sign . . . that the gains from an abundant labour saving technology receive much translation into leisure. Variations in time devoted to household obligations across our sites are not spectacularly large – roughly plus or minus thirty per cent around a mean value – and in any event correlate with household technology only very weakly. Indeed, if it were possible to take account of additional amounts of housework accomplished by paid household help (registered in our data as formal work rather than as housework) which is most prevalent in those Western countries enjoying most labour saving consumer durables, then there might well be a fully counter-intuitive relationship across the sites between the efficiency of household technology and amounts of time given over to household obligations.

This is the 'domestic labour paradox': 'labour saving devices' appear to increase the amount of domestic work.

And like most paradoxes, it may be resolved by more careful analysis of its terms. 'Labour saving devices' cannot be expected to always reduce the absolute quantity of labour – since it depends not on one but two factors; the efficiency of the equipment on one hand and the size of the task to be completed on the other. 'Labour saving devices' save labour *per unit of output*, they increase the productivity of work time. As work time becomes more productive in some particular sphere, it may be rational for people to transfer their labour resources into that sphere. Household equipment makes housework time more productive – and hence perhaps encourages people to spend more time in housework. This is a neat explanation – but it leads to a new problem: our more recent data for the UK show an absolute *decline* in domestic labour time, and data for the late 1960s and early 1970s in the USA, and the late 1970s in Holland show a similar result.[4] A contradiction on a paradox?

In fact, this does not really pose a serious problem. We have been guilty of slightly misleading the reader. The model of allocation of time described in Chapter 3 does not predict that housework time will increase with social innovation. It makes a much more strictly qualified prediction: that if the relative prices of goods and services change, and if the productivity of domestic equipment (i.e. the amount of work time necessary to produce a given amount of services with the equipment) stays constant (and if wages and preferences are also unchanged) *then* domestic work time will increase. If the productivity of the devices increases (i.e. the variable p in equation 3, Section 3.4, gets smaller) – and particularly if at the same time people become satiated with the services accruing from particular sorts of equipment – the amount of domestic work time may decline. By thinking about the problem with a little more care we can develop a quite determinate model which predicts over different historical periods, first increases in domestic work time, and then decreases.

We have some grounds for expecting the productivity of particular items of domestic equipment to develop along a specific path. We derive this from the notion of the 'product cycle'; initially, when a new piece of domestic equipment first provides a workable and attractive alternative mode of provision for a particular function, the market for the class of equipment grows quickly, and demand for it is not particularly sensitive to differences in performance between 'makes' – since *any* version of the gadget enables a considerable advance on the previous means of achieving the same end result. Under these conditions, manufacturers will concentrate on enlarging their own production facilities, confident that due to the expanding market they will be able to sell their product irrespective of the details of its performance relative to its competitors.

But as the market grows past a certain point, its rate of expansion will start to decline. As more and more households switch to the new mode of provision of the particular service, so the market for the products used in this mode approach saturation. Manufacturers are now in a more competitive situation, and they will compete with each other partly on the basis of the price of the product – and partly in terms of its performance. So there is now a motive for improving the 'productivity' of the equipment. To put the point more concretely, we might take the homely example of the washing machine. The first, top-loading electric washing-machines involved quite a lot of ancillary labour, and sold very easily, particularly as their price declined; but as the market for them approached saturation, the competition, particularly for the 'replacement market' centred around performance, and specifically around automation, the reduction of the amount of labour involved in their use. So the domestic productivity of the washing-machine initally grew slowly, but subsequently increased its rate of growth.

Now combine this pattern of evolution of the machine's productive efficiency with a diminishing marginal utility of income – in this case a specific diminishing marginal utility of the material product derived from the use of the equipment. To continue our 'clothes-washing' example; initially the increased efficiency inherent to the innovative mode of provision may have led to a very big increase in the number of clothes being washed. Perhaps the rate of increase in the size of the wash was even bigger than the increase in domestic productivity accruing from the change from the wash boiler to the electric machine – so the time spent by the household in clothes washing would increase. Or if the clean laundry had previously been acquired in the form of a purchased service, then certainly, following the Chapter 3 argument, we should expect an increase in domestic work time. But needs for clean clothes are not inexhaustible, and we might ultimately expect the rate of growth of productivity of the washing-machine (or rather the rate of decline in domestic work time required for

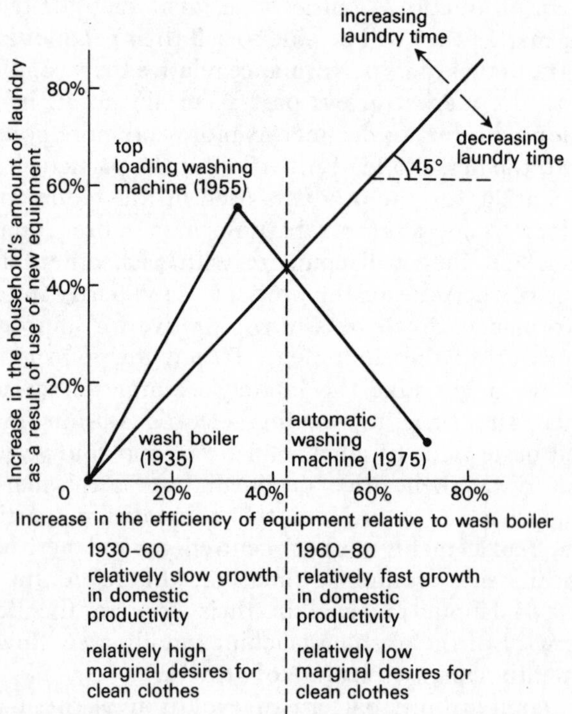

FIGURE 9.2 *Domestic output, domestic productivity, and domestic work time: the hypothetical case of the washing-machine*

a washing-machine load) to exceed the rate of growth of the quantity of clothes to be washed – so the work time decreases. Figure 9.2[5] gives a graphical interpretation of this argument.

In short, for any particular class of domestic machinery, we can expect two different processes to take place: a declining marginal utility for the particular sort of service it gives rise to; and an increasing rate of improvement in its productive efficiency. Putting the two processes together, we might draw a general conclusion as to the effect of social innovation on unpaid worktime. The change in the modal split for the particular function may initially increase domestic work time, as more households engage in the productive activity, and as the absolute output of the particular final service increases. And subsequently, as the rate of change in the society's modal split is reduced, as the efficiency of the domestic capital equipment increases, and with a diminishing marginal utility of the specific income derived from the equipment, the amount of unpaid work time may decrease.

In fact the real situation is even more complicated than this theoretical model would suggest. Figure 9.3 shows the best available estimate of the evolution of housewives' housework time in the UK since the 1930s. The time spent in housework does seem to follow the trajectory that is suggested by our 'product cycle' model – giving us the interpretation that at first (i.e. in Figure 9.3 between 1937 and 1961) output of domestic services grows faster than domestic productivity, and subsequently domestic productivity outstrips output – which would yield the initial increase and subsequent decline in the 'weighted average' curve.

But when we disaggregate the data into the two social classes, we notice a new effect. Certainly, in both cases we find the initial rise and subsequent decline of work time characteristic of the 'product cycle' explanation. But in addition we find a striking process of convergence between the classes. In the nineteen thirties it seems that middle-class housewives did something like half the working-class housewives' total of housework. By 1961 the difference between the two classes was insignificant. The middle-class housewives' increase in housework was accordingly very much faster than the working-class housewives'. What explains this difference? Presumably the explanation is not that the output of domestic services grew very much faster in middle- than in working-class households – but rather that middle-class households lost their servants. In the very steep rise in the middle-class curve over this period we have a reflection of what used to be known as 'the servant problem'.

So the full range of effects of domestic technology on housework over this period of time are probably as follows: for working-class households, the diffusion of the technology leads initially to a very substantial rate of increase in the quality and quantity of domestic services – output at first rising faster than domestic productivity, but, as the markets for the basic equipment become saturated, productivity growth gradually rises to overtake output growth; for middle-class households, output perhaps grew more slowly, but the mode of provision changed, from a predominantly 'serviced' (i.e. purchased) basis, to a 'self-serviced basis'. As we might expect, we have a combination of Engel's Law and social innovation effects; working-class households acquiring more of the 'luxury' domestic services –provided through the innovative mode; middle-class households experiencing less of an increase in provision for this category of final service function, but nevertheless changing their previous mode of provision for it.

9.3 The Distribution of Domestic Work

This section discusses two different distributional issues; the *segregation* of male and female work tasks, and the *distribution of total household work time*. There are two reasons for discussing these issues separately. First, there is a methodological problem; it seems likely that the 'diary' method used in time budget studies may overestimate paid work time relative to unpaid (since snack breaks, conversation time and similar activities tend to be recorded less faithfully at the workplace than in the home) – which in turn leads to an overestimate of men's work time relative to women's. The evidence we get from time budget studies about the *segregation* of tasks is much less subject to question. Second, the indisputable evidence about segregation is in itself sufficient to suggest the need for redistribution of domestic work responsibilities. Nevertheless, in spite of its shortcomings, the 'distribution of total household work time' data is significant – we shall turn to consider it in a moment.

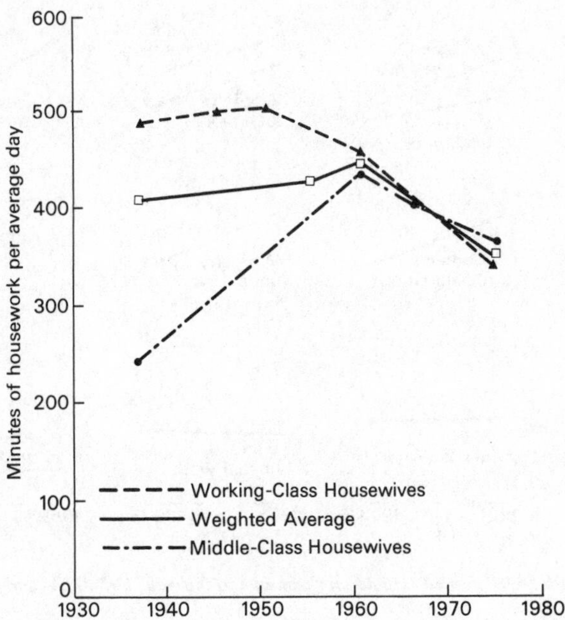

FIGURE 9.3 *Housewives' domestic work (non-employed, or part-time-employed women)*

It appears that women are subject to a prescription that they should bear the major part of their household's domestic work and this prescription is sufficient to maintain the unequal position of women in paid work. Fig. 9.3 compares the paid and unpaid work data for the UK, USA and Japan. In the UK and the USA women work on average slightly shorter total hours than men do, while in Japan they work considerably (more than 20 per cent) longer hours. But what all three countries have in common is that women bear a disproportionate (and, across countries, all in all, surprisingly similar) responsibility for unpaid work. Fig. 9.5 makes the point more specifically. Having children increases the work total for both the sexes; and indeed, it increases in all the cases, the *domestic* work. But women's work is increased more than men's. And, most important for the argument, while men's *paid* work is hardly affected at all, women's paid work is very considerably reduced (by 38 per cent in 1961 and 35 per cent in 1974/5).

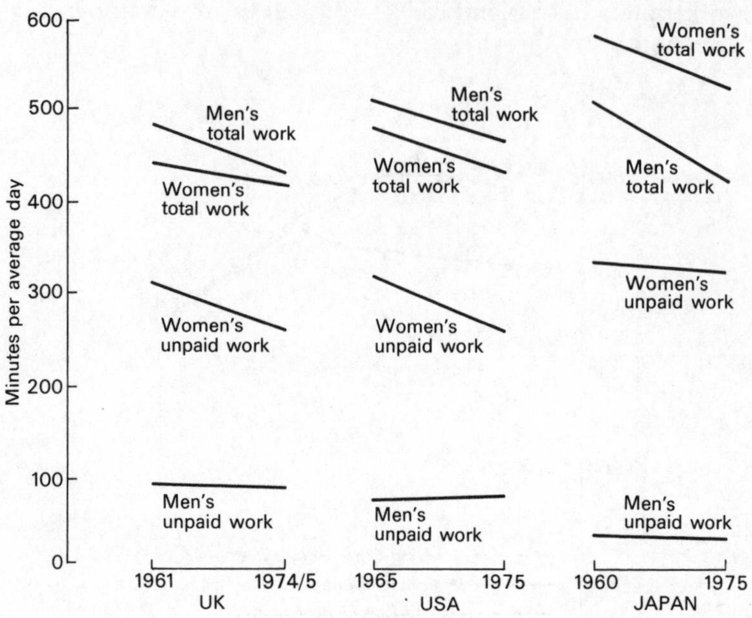

FIGURE 9.4 *Paid and unpaid work, men and women, UK, USA, and Japan*

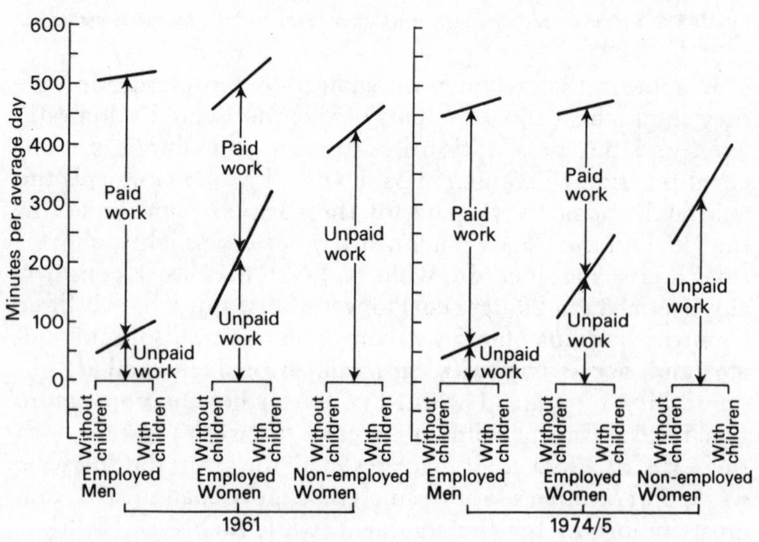

FIGURE 9.5 *Total work by sex, paid employment, and family status (T-statistics, UK aged* <*45)*

Irrespective of any inequality in the total hours of work, an inequality is certainly to be found in the *composition* of work time for the sexes. Men tend to spend about one-fifth or one-sixth of their total work time in domestic work, women, about one-half. Whether or not this is at any level a matter of choice, either by men or by women, this is an enormous difference, far greater than could be explained by the biological requirements of motherhood. And of course this is *not* a matter of choice. Despite recent and quite substantial changes in publicly expressable attitudes, domestic work is still regarded as chiefly women's work. The fact of their perceived responsibility for domestic work reduces the amount of time women have available for paid work, *which places them at a disadvantage in the wage labour market.* And here is the inequality: precisely *because* similar sorts of men and women, *because* husbands and wives may be expected to want to work similar lengths of time, while women maintain their special responsibility for housework, women are more likely to take on part-time jobs (which are generally of inferior status), they have less energy to concentrate on their jobs, less flexibility to work longer hours when they are needed, so they cannot compete on equal terms with men.

And in spite of what we might consider to be a quite clear requirement of sexual equity that the total work should be approximately equally divided between men and women, it does seem clear that this requirement is not met. Table 9.2 shows that women in certain sorts of household (and men in others) work on average considerably longer hours than their spouses do. Table 9.3, which calculates the male proportion of the spouses' total work, makes the point clearly. Remembering that the husbands' proportion of total work is probably overestimated (by perhaps 3–5 per cent) because of the inclusion of rest-breaks in paid working time, we can see a certain symmetry. In households with children where the husband has a paid job and the wife is non-employed, the spouses probably have just about the same total of work time. And if there are no children in such households, the husband may do rather more work than the wife (though we should recall from Figure 8.5 that in most cases where there are no children in the household the wife does in fact have a job). But where the wife does have a job, and particularly where she has a job

TABLE 9.2 *Work times in various sorts of households, UK 1974/5*

	Husband and Wife Employed Full-Time			Husband Employed Full-Time Wife Employed Part-Time			Husband Employed Full-Time Wife Non-Employed		
	Younger, No Children	Children in Household	Older, No Children	Younger, No Children	Children in Household	Older, No Children	Younger, No Children	Children in Household	Older, No Children
Number in Sample	42	95	22	6	164	28	17	234	41
Minutes per Average Day									
Husband: Paid Work	382	390	383	463	404	401	417	370	373
Unpaid Work	106	101	76	75	84	68	75	99	81
Total Work	488	491	459	538	488	469	492	469	454
Wife: Paid Work	339	304	354	286	172	179	26	6	–
Unpaid Work	147	216	194	171	296	267	290	404	339
Total Work	486	520	548	457	468	446	316	410	339
Couple: Paid Work	721	694	737	749	576	580	443	376	373
Unpaid Work	253	317	270	246	380	335	365	503	420
Total Work	974	1011	1007	995	956	915	808	879	793

and children, she appears to bear a disproportionate burden of the household's total work.

TABLE 9.3 *Husbands' proportion of couple's total work time, UK 1974/5 (Overestimating husbands' proportion by 3–5%: see text)*

%	Younger, No Children	Children in the Household	Older, No Children
Both husband and wife employed full-time	50.1	48.6	45.6
Husband employed full-time, wife employed part-time	54.1	51.0	51.3
Husband employed full-time, wife non-employed	60.9	53.4	57.3

It is clear from Table 9.2 that the reason for this disproportionate burden is that employed wives tend to do their jobs *and* the housework; one part of the household's work – the part of the formal economy – is desegregated, while the other part – in the household – remains a segregated female activity. Employed wives admittedly do less housework than non-employed wives – but the reduction is nothing like proportionate to their increase in paid work; so full-time employed women work something like two hours per day longer than non-employed women in equivalent family circumstances. And husbands with employed wives, though they do increase their unpaid work a little, do not do so by anything approaching the amount that would be necessary to bring total work times to equality.

It appears that the traditional segregation of domestic tasks means that as long as the equivalent traditional pattern of segregation in paid employment holds, the equitable requirements, that the total work times of the spouses should be roughly similar, is met. But when the segregation of the paid work-force breaks down, there is no equivalent breakdown of the pattern of segregation in domestic labour – leading, in some sorts of households, to quite gross inequalities.

However, the conclusions from the time budget data are not entirely uncomfortable. It does appear that, over time, some of the disparities may be declining. Figure 9.4 shows that the disparities between employed men with children and employed women with children may have decreased in the UK between 1961 and 1974/5. In particular the additional amount of work caused by children for employed women seems to be less at the later date than the earlier – perhaps as a result of the sort of technical changes described in Section 9.2. And the male proportion of the household's total domestic work appears, from Table 9.2 to be increasing – though more as a result of women doing less domestic work than of men doing more.

One more general observation emerges from this argument. We have an essentially *sociological* explanation for the development of the division of domestic labour; we have explained it, in effect, by people's perception of their roles. We have women, entering employment without appropriate role models, attempting to fulfil the roles of housewives in addition to that of employee, while their husbands continue to act out the traditional role of bread-winner even though they are no longer married to traditional housewives. If the households were behaving rationally, they would presumably go through a process of comprehensive reallocation of tasks, with both spouses reconsidering their responsibilities for housework when the wife reconsiders her responsibility for paid work. Or, to be more analytically precise, we have two possible conclusions from the data we have considered; either the household does in fact reconsider the distribution of all work activities, in the way some economists have proposed, in which case (given the inequality in total work hours) we must conclude that husbands are in general in a position to exploit their wives; or else, which seems very much more likely, they do not allocate tasks according to any economic rationality but simply according to traditional patterns, without adjusting the distribution of some tasks even in response to quite big changes in the distribution of others.

9.4 Summary of Chapters 8 and 9

Some rather clear findings emerge from this and the previous chapter:

1. It appears that virtually everybody participates regularly in both sorts of work during at least part of their lives. Within

households, the replacement of purchased services by 'self-service' production means that even the rich, who might previously have avoided it, become involved in some form of non-specialized domestic work. Virtually all men show at least some token participation in this sort of activity. As regards participation in the money economy, the great majority of men have full-time jobs for most of their adult lives, and an increasing proportion of women have some sort of participation for at least part of their adulthood – though most of the more recent increase has been in part-time jobs. Women also show an increasing tendency to *return* to work after a period out of the paid work-force during which they have been mainly engaged in child-care. This period spent out of the paid work-force (or in part-time work) is much longer than that during which children are biologically dependent on their mothers. Even women with young school-age children (5–9) show markedly lower participation in paid work (and in full-time paid work) than women with older dependent children.

2. While participation (P-) statistics for both sorts of work have been increasing, the participants' time (T′-) statistics have been decreasing. In paid work, hours of manual and non-manual, men and women, seem to have been declining for a long period – certainly since the 1950s, also, though the statistics have not been presented here, excluding a trough in the 1940s, for a long while before. Evidence, both on manual and on other hourly paid workers suggests that the decline may be interpreted as evidence of a diminishing marginal utility of income; a similar phenomenon presumably applies to non-manual workers. The T′-statistics for unpaid work over the last two decades show a decline, which contrasts with the constant or rising trends found (in USA studies) for the 1930s–1960s. It may be that initially the high productivity of 'labour saving' (per unit of output) domestic gadgets may have led to such a large increase in the total outputs of households that the total work done in households absolutely increased; whereas later continuing increases in domestic productivity did not lead to as much output growth so that labour time declined. Fancifully, perhaps, we might interpret this phenomenon as evidence of a diminishing marginal utility of *non-money* income.

However we may speculate that the rising trend in the USA reflects, as the UK evidence suggests, the disappearance of domestic servants from middle class households.

3. In fact, if we combine the P- and T′- statistics into T-statistics, for paid and unpaid work, we can advance a rather general hypothesis of DMUI for the total income (i.e. money plus non-money) of socieities. Figure 9.5 stakes the T-statistics (paid plus unpaid work) for the various countries which participated in the multinational time-budget study organized by UNESCO, plotted against their 1970 National Income. There does appear to be a relationship; past a certain level of economic development, total social hours of work seem to decline. But, of course, cross-sectional evidence of this sort is not really reliable, particularly as there are systematic political differences between the industrializing and the industrialized group.But we do also have some longitudinal data for some countries, and, as we see from Figure 9.6, these do conform rather well to the cross-sectional picture.

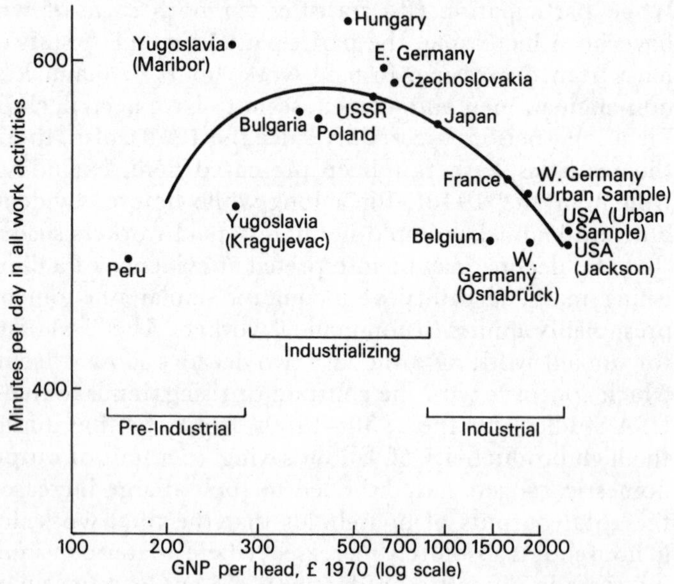

FIGURE 9.6 *Total hours of work and economic development*

On balance, it would seem sensible to conclude from the evidence that a resumption of economic growth would lead

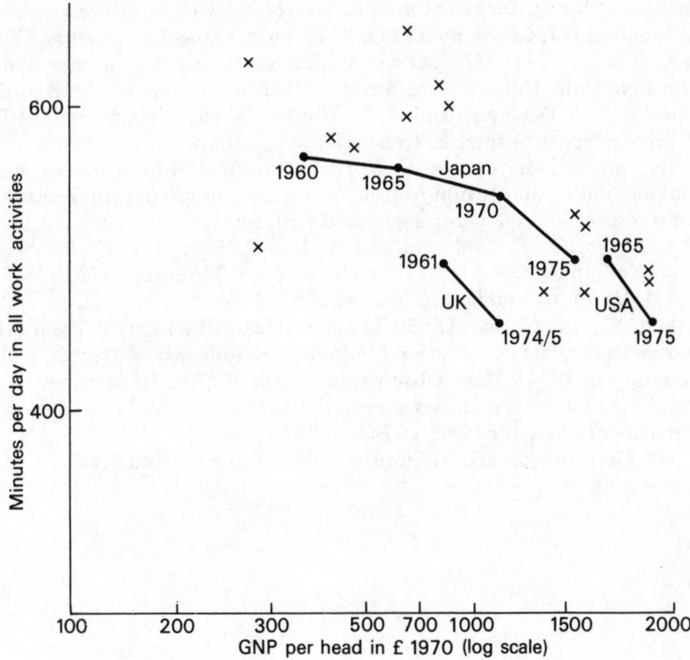

FIGURE 9.7 *Change in total work time: UK, USA, and Japan*

to a continuation in the decline of working hours. Emerging from the DMUI conclusion is a clear implication for the problem of unemployment. The resumption of economic growth would help paid employment in two ways. It would obviously, provide more demand for labour. It would also (via DMUI) reduce the labour supply in the sense that, even if the participation *rate* were to continue to rise, participants' *time* in employment would be expected to fall.

NOTES

1 'Time-budget' studies involve the collection and analysis of detailed daily or, as in our case, weekly diaries which specify the timing of an individual or household's activities. Our study involved the reprocessing of two national random-sampled sets of diaries (2,500 diaries in 1961, 3,500 in 1974/5), collected by the BBC Audience Research Department, originally for the purpose of assessing the availability of viewers and

listeners at home. Details of the procedures involved in this research will be found in J. I. Gershuny and G. S. Thomas, *Changing Patterns of Time Use, U.K. 1961–1974/5*, SPRU Occasional Paper 13, Science Policy Research Unit, University of Sussex, 1980; an outline of the results is found in J. I. Gershuny and G. S. Thomas, *Social Change and the Use of Time*, Report to the UK Sports Council, 1981.

[2] By 'unpaid' or 'domestic' work in this section we mean routine cooking and cleaning, non-routine 'odd jobs', shopping, child-care, and the transport and queueing associated with these.

[3] J. Robinson, P. Converse and A. Szalai, 'Everyday life in Twelve Countries', in A. Szalai (ed.), *The Use of Time*, Mouton, 1972, p. 125.

[4] J. I. Gershuny and G. S. Thomas, 1981, Table 3.6.

[5] J. I. Gershuny and G. S. Thomas, Mass Observation Manuscript Diaries 1937; Mass Observation Limited 'The Housewives' Day' (working class only), 1951; Mass Observation Limited 'The Housewives' Day' (middle and working classes averaged), 1956; BBC Audience Research Department data, 1961 and 1974/5, 1983.

[6] J. I. Gershuny and G. S. Thomas, 1981, Table 6.3 and 6.4.

[7] Ibid.

CHAPTER 10

Two Waves of Innovation

10.1 The Argument in Brief

The foregoing arguments suggest that we must replace the conventional explanation of change in economic structure with a slightly more complex, but perhaps ultimately more insightful one, which might be summarized as follows:

> As societies become richer, their pattern of 'needs' (for 'final service functions') changes; they may demand more 'luxuries' relative to basic commodities. Over time, techniques for satisfying needs may change ('social innovation'), so that increasing demand for luxury commodities does not necessarily imply increasing demand for final services, and may indeed be for manufactures. As the technology of production processes advances, however, production tasks become more specialized and 'fragmented', and as a consequence we may expect a growth in intermediate demand for services (and also a growth in demand for more specialized workers in 'service occupations' throughout the economy). Service industries show generally lower rates of productivity growth than those of other industries; so employment in services may grow as a proportion of the total even if there is no proportional growth in demand. But the existence of this 'productivity gap' between services, and other industries may actually encourage changes in techniques for satisfying needs, thus reducing final demand for services. The change in sectoral distribution of employment over any period is dependent on the balance amongst these various processes.

The first sections of this chapter will consider two periods; 1950–80, in the light of the evidence presented in Chapter 6 and 7; and, speculatively, 1980–2000. This latter, futurological, exercise is intended merely to give a picture of what may technically be possible, as opposed to what actually may be the course of economic events in the 1980s and 1990s. The final sections of the Chapter contain a more general view of some possible alternative futures – and a description of the new model of the development of developed societies that has emerged in the course of the arguments of this book.

10.2 The 1950s Wave

We can put together a rather simplified, schematic view of changes over the period 1950–80 using the structure of the model developed in Chapter 6. Let us start, as we did in Chapter 6, with final consumption. We may certainly guess that the trends found in Table 6.4 are more marked over the longer period. It seems clear that, at least from the mid-1950s, the proportion of final consumption accounted for by the basic shelter and food functions has been declining in favour of domestic services, entertainment, transport, education, and medicine, with the latter two becoming of dominant importance towards the final decade of the period.

But changes of particular importance for the development of European economies have taken place *within* some of these functions. There has been substantial social innovation in domestic services, entertainment, and transport. In domestic services, this reflects consumer durables such as washing-machines and vacuum cleaners, combined with domestic electricity supply; in entertainment, it reflects equipment such as television sets, together with broadcasting infrastructure and 'software storage media' such as records and magnetic tapes; in transport, cars, and road networks. In all these areas, the modal split has changed, giving smaller proportions of expenditure on marketed final services (e.g. domestic help, cinemas, public transport) and a larger proportion on goods. Through the period public provision has been of growing importance in medical and educational functions – and presumably because of the complexity of the functions in relation to the available technology there has been little significant social innovation here. The net consequence of the shifts in the functional distribution and the changing modal split within some functions has probably been to increase the proportion of final expenditure devoted to non-service commodities, to diminish the proportion devoted to final marketed services and to increase non-market service consumption.

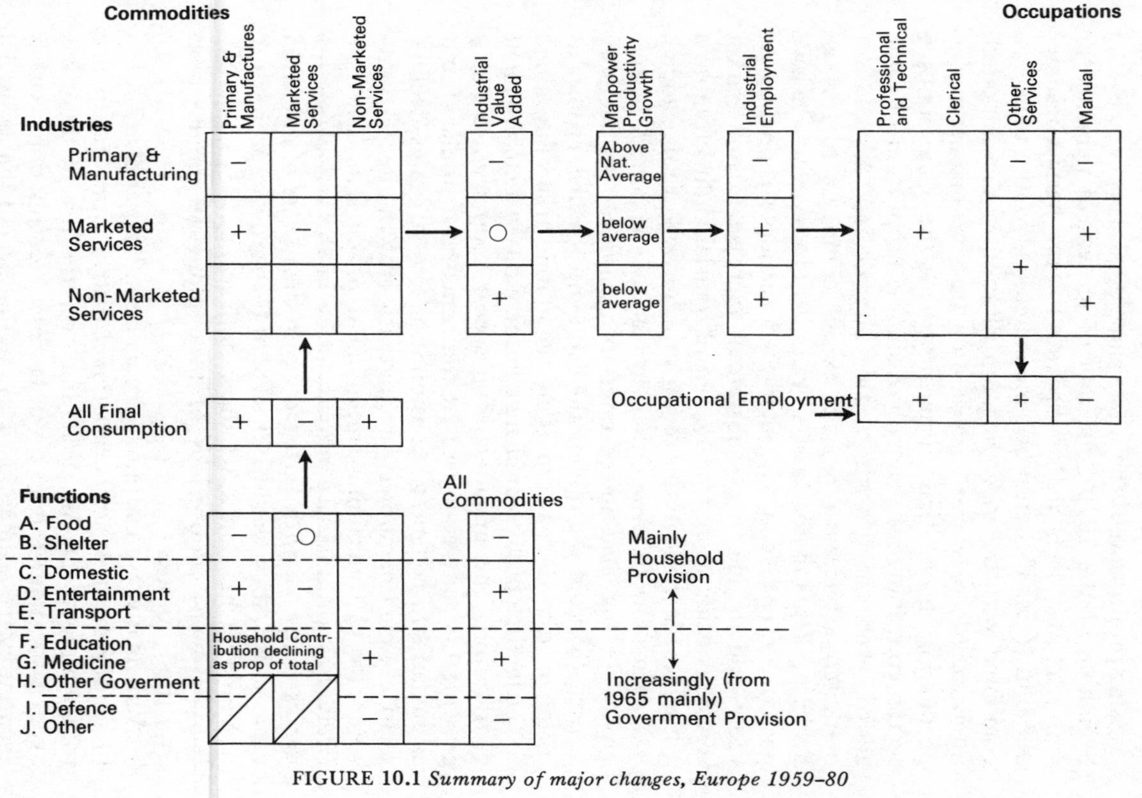

FIGURE 10.1 *Summary of major changes, Europe 1959–80*

In the first half of the three-decade period, the increase in demand for manufactured goods had the consequence that in many countries, the manufacturing sector contributed an increasing proportion of the total value added. However, by the beginning of the 1970s, the rate of growth of demand for manufactures slackened as markets for the basic consumer durables (cars, TVs etc.) became increasingly saturated. This, combined with the increasing contribution of the marketed, intermediate, *producer service* industries to manufacturing production, meant that during the 1970s the manufacturing sector contributed a generally declining proportion of value added. And the growth of this intermediate demand for marketed services to a great extent compensated for the declining demand for final services, to give no very decisive pattern of change in the marketed services proportion of value added. The very substantial increase in non-market service consumption over the period translated directly into a quite clear increase in this sector's proportion of value added.

It seems safe to assume that throughout the period the two service sectors had lower rates of manpower productivity growth than did the rest of the economy – which enables us to move from the pattern of change in industrial value added to the familiar pattern of change in the industrial employment distribution: the manufacturing sector proportion initially growing in many countries (at the expense of primary), but as manufactured goods' consumption growth rates decline, relatively high manpower productivity growth rates lead subsequently to a declining employment proportion. The relatively low productivity growth rates in the service sectors means that even with zero growth in their share of industrial value added, they would still account for a growing proportion of employment – and as we know, both accounted for substantially growing proportions of total industrial employment over these decades.

And finally, we can move from the industrial distribution of employment to an *occupational* distribution, and thence to *un*employment. We find that over the period there has been a substantial shift towards service occupations within each industrial sector; that is to say that even if the industrial distribution of employment had remained constant, there would still have been a very substantial growth in the service occupations. We might estimate that perhaps half of all growth in service occupations may be attributable to intra-sector rather than inter-sector shifts. And as, during the 1970s, overall manpower productivity growth has gradually come to outstrip growth in output, unemployment grows and, as we showed in the previous chapter, unemployment rates for particular occupations rise in inverse proportions to the growth in demand for these occupations.

Viewed from this angle, these trends would lead us to rather pessimistic prognosis for the future of employment. For manufacturing, the 1950s wave of innovation seems to be just about played out; the few new products that have emerged (e.g. video recorders) do not seem to have any very great employment generation potential – and in general the scope for substantial new social innovations with existing infrastructure seems minimal. Certainly we have nothing in immediate prospect to rival the growth opportunities afforded by cars and domestic equipment in the 1950s and 1960s. For non-marketed services, decades of relatively poor performance in productivity growth seems to have resulted in a general public perception that government expenditure is likely to be wasteful (and in particular, that expenditure increases are likely to be translated into wage rises rather than enhancements in performance). So non-inflationary increases in government expenditure are not in general likely. The decline in demand for final marketed services can only be reversed by altering the trend in service costs, which means in practice, cutting wages. In sum, simple extrapolation from our experience over the last three decades yields only the prospect of declining wages and rising unemployment.

10.3 A 1980s Wave of Social Innovation?

However the theoretical apparatus we have developed can be put to more positive use. It appears that there may now be at least the technological potential for a new wave of social innovation. This new wave would employ microprocessors and cheap information storage devices, together with a new information-transmission infrastructure combining the flexible linkages of the telephone system with the broad band-width available on cable television systems. It would consist of innovations both in the domestic, entertainment, and transport/communications functions affected by the 1950s wave, and, in addition, in education and medicine.

We can list some examples of innovations which have already been proposed (in many cases these are already at a stage of working demonstrations):

- In *domestic services*, we can construct systems for automatic centralized monitoring and control of a range of household functions (heating, lighting, safety); and these household systems can themselves be linked to local security or safety services. In addition we might imagine information packages giving advice on household operations.
- In *entertainment*, there are already operational 'home-box-office' and analogous systems, giving households the option of access (via cables) to a much wider and more varied range of entertainment material than could be provided on a mass broadcast basis. By a simple extension, we could imagine subscription schemes, using such systems to promote new films, plays, musical performances which would otherwise not find a market either in theatres or in broadcasting.
- In the *transport/communications* area, the same infrastructure and domestic equipment could enable electronic funds transfer and 'remote shopping'. There would also presumably be facilities available for video telephones and other sophistications of the current telephone system (such as computed switching to enable a conversation not with any *one specified* individual, but with *anyone* who wished to talk about some specified topic).
- In *education*, we could imagine the proliferation of packages for remote, and if necessary interactive education or training.
- In *medicine*, we might foresee continuous remote monitoring of chronic disorders, enabling more home care. Similarly we might perhaps develop systems for remote diagnosis – and certainly interactive packages for medical and other counselling would be a sensible use of the available facilities.

Considering each of these as *individual* innovations, certainly the costs – particularly the infrastructure costs – would be

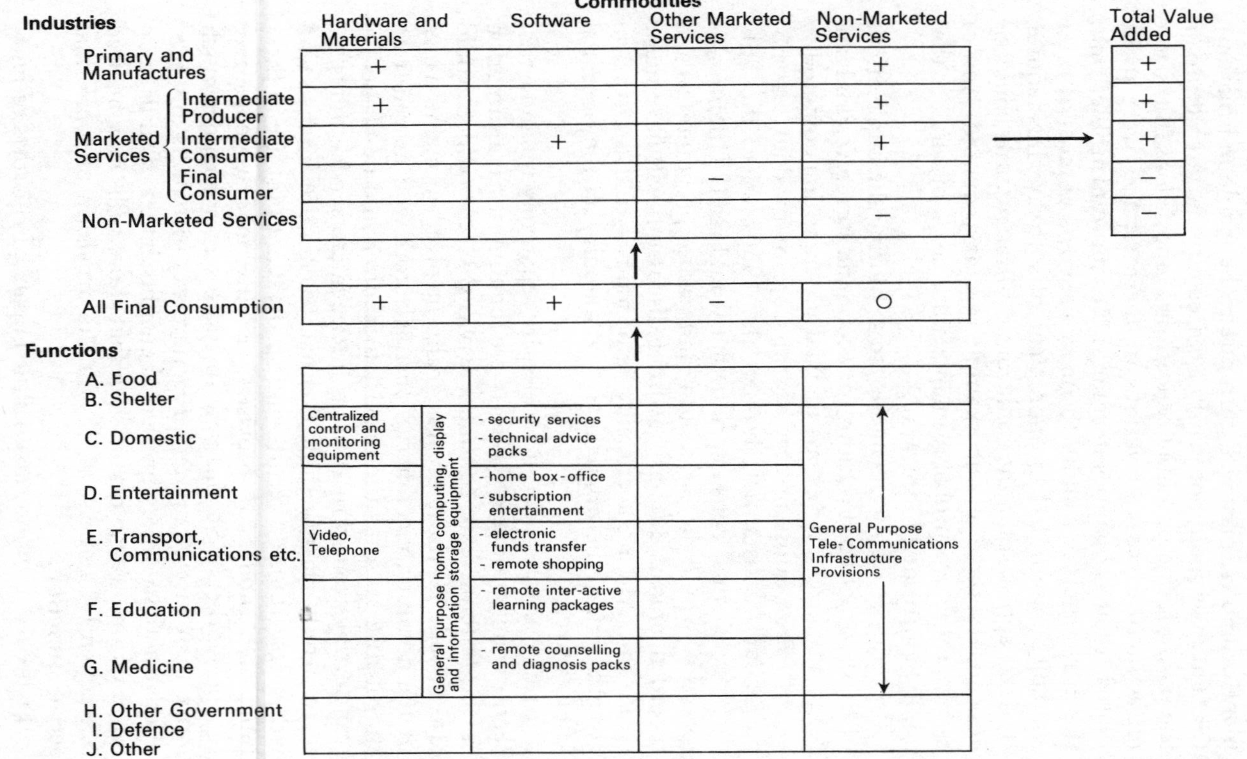

FIGURE 10.2 *Technologically feasible changes in households' final consumption patterns and their consequences for industrial value added*

prohibitively high (though some, particularly the entertainment examples, may well turn out to be viable in isolation). Certainly we could not justify a 'wired city' infrastructure on the basis of, say, potential educational innovations. But if the infrastructure costs of various of these services were to be shared, their economics become rather more plausible. Once the infrastructure is built, the marginal costs of most of these examples amount to little more than the software they require. It does appear that the same basic infrastructure could be designed to serve all the examples we have quoted. And of course there are probably many more sorts of services that could be provided in this way.

In order to apply our model to exploring the implications of these sorts of innovations for economic structure, it is helpful to think of a new class of final consumption commodity – *software*. Consumption of this sort of product was not particularly substantial through the 1950s wave; the only category of final consumption that would really fall into this category would be television and radio programmes and recorded music. But, as our brief discussion will have made clear, this will be a crucially important – and perhaps, in growth terms, the *dominant* – commodity in the potential 1980s wave.

It may also be helpful to identify separately the industrial sector which produces this software for domestic consumption as the 'intermediate consumer services' sector (which would also include all those professionally involved in maintenance of the domestic machinery). Just as actors who moved from the theatre stage to the television studio could be classed as transferring from a final to an intermediate service industry, so we would classify those doctors, teachers, or firemen who write algorithms or record video programmes for use in these innovative modes of provision as 'intermediate consumer service workers'. So this new software sector includes much more than just computer programmers – it covers any productive activity which involves the embodiment of skills in an information storage device such that those skills may be used subsequently, and elsewhere, for the provision of some final service function.

Figure 10.2 shows the possible impact of the sorts of innovations we have outlined for the structure of industrial value added. We would expect the innovations, initially at least, to stimulate manufacturing production (just as in the first part of the 1950s wave). Certainly there would be very considerable growth in the construction industry during the initial installation of the infrastructure. Both sorts of intermediate service industry might be assumed to benefit, while the marketed final consumer sector would have an accelerated rate of decline. Even if government consumption expenditure were to rise, in this model most of the growth would be concentrated in manufacturing, construction, and the intermediate services, so value added in the non-marketed service sector itself would also probably decline.

This is of course the very vaguest of possible outlines. But we must stop here: so far we are only discussing technical possibilities. To see whether this optimistic picture is realizable, or whether the pessimistic extrapolation of the previous section is more realistic, we must consider the broader range of social, political, and economic issues which underlie the choice of techniques.

10.4 The Productivity Gap and the Welfare State

The examples discussed in the previous section relate mainly to households' expenditure. But these domestic innovations may well find parallels in the provision of public, non-marketed services. We should at this point make explicit the reason for thinking that innovation in public provisions must move in line with private.

As a general rule, we might assume that if two sectors have different rates of labour productivity growth, the relative prices of their products will change. As we suggested previously, if the 'labour share' stays constant, and if people with particular sorts of training, or in particular occupations, in the low productivity growth sector (services), wish to maintain parity (or existing differentials) with equivalent workers in the high productivity growth sector (manufacturing), the conclusion follows inevitably. As we have seen, such a 'productivity gap' is a major explanation of social innovation in a number of final service functions; the changing relative prices lead to a change from the household purchase of final services to the purchase of goods.

But this flexibility is not possible in the acquisition of public goods. Expenditure on public services is forced through taxation; public services serve a population of captive consumers. Only the richest households can afford to ignore the non-marketed services offered to them, and purchase privately produced alternatives, however unsatisfactory they may find the public provision. And such action in any case relates to the perceived quality of the provision not the price – private purchasers of educational and medical services, in countries where these are provided by the state, are paying twice over for them. So the flexibility, the adjustment, enabled by social innovation, and driven by market forces in private consumption, is not available in public provisions.

In a growing economy, this may not have any particularly unpleasant consequences. True, even a constant level of provision of public services will increase in cost over time, as wages, but not productivity, rise in the public sector in line with the private. But since education and medicine are relatively high in our hierarchy of needs, people may be quite happy to increase their absolute expenditure – and even the proportion of national income – devoted to these provisions, as long as this does not prejudice other real increases in levels of provision for other functions.

Consider, however, the implications of the productivity gap in an economy without any real growth in national product – but which nevertheless has, as a result of technical change, continuing labour productivity growth in its manufacturing sector. The absolute level of GNP is constant; if, on previous arguments, wages in the service sector keep in step with those in manufacturing, the proportion of GNP necessary to maintain a constant level of Welfare State provision must increase, and so real consumption of other commodities must be declining. In such a situation, unemployment will probably be rising. So the maintenance of the welfare state is associated with a reduction in the level of non-service sector consumption – probably by reducing the real incomes of the unemployed.

Of course, there is a certain economic sleight-of-hand here. We have simply defined a set of circumstances; there is no influence to be drawn that redistribution is a consequence of the service provisions (though of course there are economists who would argue that this is in fact so). But nevertheless the argument does point to a problem facing governments in zero-growth economies. They must choose between allocating income to public service workers or the unemployed. The growing cost of the steady-state public welfare service does not represent any additional provisions of services, but merely its growing wage bill, with the result that the extra expenditure may be difficult to justify to an electorate – the lack of productivity growth, in short, is subversive to the continued existence of the welfare state. And one very frequent consequence of the conflicting pressures is a reduction of investment expenditure in the Welfare State – with further adverse consequences for its productivity.

So there are two reasons why we should look for innovations which increase the productive efficiency – increase the output per unit of expenditure – of non-marketed services. The first is that if we do not do so, then we are cheating the people who rely on the public services, who are entitled to the best possible services in return for the taxes they pay. And second, where public expenditure comes under especial scrutiny, the rising cost of a constant level of provision may be difficult to justify, so the existence of the Welfare State is itself threatened. It is admittedly very difficult to measure exactly what the output of the non-marketed service sector is – but this difficulty is no valid reason for dissuading us from trying to increase its output relative to its inputs.

So: what sorts of social innovations must we think about for the public services? The ground rules for innovation here are, in general, the same as those for private innovations; embodying skills, providing more capital, using informal labour. Let us, to start with, consider an example that involves no technical, but only organizational change.

Consider the provision of crèche facilities for very small children. At present, some (though very few) local authorities in the UK provide such facilities; let us assume that, as in most services, something like 70 per cent of the total costs of these provisions is accounted for by labour, and 30 per cent by capital equipment. At present these services are provided on a fully staffed basis; suppose the local authorities were to go over to a mode of provision of crèche services that involved a combination of voluntary labour (parents) fulfilling the primary function of caring for the children – working perhaps one hour for every five that the child is in the crèche – with professional paid labour providing technical, advisory, and organizational skills. Let us assume that initially the overall level of expenditure remains constant, but the proportional allocation of costs switches to 70 per cent on capital and 30 per cent on paid labour. For the same public expenditure, we would get nearly two-and-a-half times as many crèche services (assuming the capital equipment has the same efficiency in both modes of provision).

Under the assumption of constant expenditure, paid employment would fall, to a little less than half its previous level. But this assumption is no longer a necessary one. The provision of services has become very much more efficient; it becomes rational for the local authority to transfer expenditure in this sphere of service provision – by very little more than doubling its expenditure, it can increase its output of crèche service fivefold *and* maintain the level of employment. And, to generalize from the example, if all public services were to increase their productivity, it would be rational for households to wish to transfer expenditure from disposable income to their taxes to pay for them. Just as demand for services shows a downwards price elasticity with rising real prices, so it would presumably show an upward elasticity if effective prices were falling; public expenditure may be encouraged to increase by productivity growth.

In general, the public service sector innovations may be expected to have a technological component as well as an organizational. Our second example comes from the opposite end of the educational spectrum. Let us consider the case of the UK 'Open University' as compared with a conventional university. The conventional university has relatively low fixed costs – mainly buildings, libraries, laboratories – and high variable costs, since additional students require additional staff. The OU has a quite different cost structure, since its teaching is based mainly on radio and television broadcasts and postal material. It requires a considerable fixed expenditure – in the preparation of 'course units' and in broadcasting them. But its variable costs are very much lower than the conventional university, consisting mostly of postage and local tutorial services. Above a certain level of output (around 1,500 students per course) therefore, the OU provides a

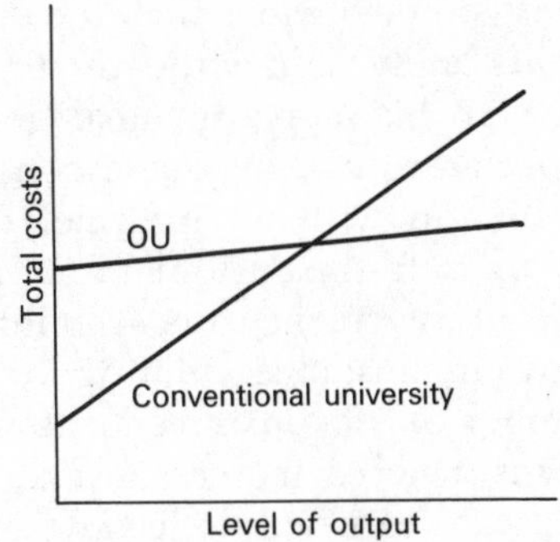

FIGURE 10.3 *Conventional vs Open University provisions*

cheaper service than a conventional university. (On average the direct cost of an OU degree is about one third of a conventional one.) It achieves this by using less labour and more capital in the formal economy and using household facilities and presumably some domestic labour in the final production of the final education service.

This is not, of course, to suggest that the Open University is an appropriate basis for the redesign of all higher and further education. But consider the effect of the application of some of the OU techniques in conventional universities. The OU course material is of the very highest quality – very much better than the average standard of teaching in British universities. This is a consequence of the substantial resources available for the preparation of the courses. While a teacher in a conventional university might have at best a few hours available for the preparation of a lecture, an equivalent presentation in the OU might receive a person-month or more of preparation, from the specialist teacher working together with research assistants and media experts. Particularly for the teaching of the more standard undergraduate courses, both teachers and students could benefit from a pooling of resources (over time and between institutions) to prepare recorded material, and interactive computer-based teaching systems. The teachers would have more time to prepare the most appropriate and up-to-date material (assuming the course units were continuously revised); the students would have better constructed courses – and more time might be freed for direct teacher–student contact.

The implications go much wider than the university sector. The same arguments apply right across the tertiary education field. And there may be scope for analogous innovations, particularly in specialist teaching, in secondary schools – perhaps in combination with an informal contribution of communal labour to fulfil the custodial functions of the educational system. As in the crèche case, such innovations *could* (and under present circumstances admittedly would) be used to reduce the number of jobs in the educational system. But this is by no means a necessary conclusion. Needs for education are by no means satisfied; indeed we might argue that they are in general frustrated by the difficulties of providing substantial increases in educational facilities by marginal increases in expenditure. It may be that the increased demand for education that might be generated by such innovations could in the longer term lead to maintenance or even an increase in the size of the educational establishment.

Very similar arguments could be developed for the medical sector. Consider for example the effect of a universally available computerized initial diagnosis and screening system which directed prospective patients to an appropriate medical or paramedical specialist (or recommended self-medication) – to replace the General Practitioner who currently fulfils this function. Perhaps half of those who currently visit the GP would not then do so. This could mean that fewer GPs would be needed – or it could mean that each GP could spend longer with each patient, and provide a higher level of service. Or consider the combination of voluntary provision of the catering and room-service functions of hospitals with sophisticated monitoring and advice systems and specialist medical labour . . .

The point is that such, or similar, innovations are possible, and if they are possible and yet are not taken up, we may be providing fewer services than the public expenditure could support – cheating the service recipients, and, particularly, cheating the poor, who have no access to other than the public provisions. And by not innovating we may be threatening the welfare state itself. Jobs in public services may initially be at risk: but after all, public services exist to provide the public with services, not to provide service workers with jobs. And in any case, such radical improvements in the effectiveness of provision might in the longer term enable markedly increased expenditures and growing employment. Simply: non-marketed services, like any other industrial sector, must ultimately compete for custom against the alternative ways that people might wish to spend their income. Its employees have no unquestionable right to employment.

10.5 Three Mechanisms for Generating Employment

To complete the theoretical argument, we must consider the three ways that jobs might be generated if our new 'wave of development' were to get under way. The first is the conventional, direct effect of economic growth, but the other two are less usually considered, and arise out of our discussion of the labour supply in Chapters 8 and 9.

If the economy's output grows faster than its labour productivity, then (almost, but as we shall see not entirely) necessarily, its employment increases. Certainly there is a positive correlation between these two variables – an increase in output itself has the effect of increasing productivity – so that, given a small but positive rate of productivity growth, even with zero growth in GNP (which leads to rising unemployment), it takes perhaps 2 or 3 per cent output growth before unemployment begins to fall. Nevertheless this is the main mechanism to which most people look for reducing unemployment; and it is also the main reason for pessimism over the medium term prospects for employment in the UK. After all, if we do have to have annual GNP growth rates of something like 3 per cent to reduce unemployment at all, then it might take more than a decade of sustained output growth at the implausibly high level of 4 per cent (which is higher than anything we achieved in the 1950s or 1960s) to reduce unemployment to its mid-1960s level.

But fortunately there are two other mechanisms which might reduce the size of the problem. Let us be clear that job-sharing, in the sense of people giving up a proportion of their work time and also giving up a proportion of their income, in order to provide other people with jobs, is quite implausible. This would assume that people have, not a *diminishing* marginal utility of income, but a *negative* marginal utility of income, a proposition for which empirical evidence is quite lacking. Though people working a particular work week may tend to value the income from the next additional hour of work below the benefits of their marginal hour of leisure, they tend not to be willing to forgo the last hour of work if they must thereby forgo an hour's pay. They may, as we argued in Chapter 8, be willing to take a proportion of the society's *extra* productive potential in the form of leisure, but not to absolutely reduce their income in return for an increase in leisure.

This observation does however give a clue to the second way in which economic growth may generate jobs. Consider the evidence discussed in Chapter 8 concerning the approximately 11 per cent reduction in average working hours between 1955 and 1975. What would have happened if over these two decades British output and productivity growth rates had been as they actually were, but average hours of work had stayed constant. Without wishing to fall too heavily into what has been termed the 'lump of labour fallacy' (i.e. the assumption that each hour of work in the economy is equivalent to each other hour) there is some sense in observing that unemployment would have been 11 per cent higher – $2\frac{1}{4}$ m. more unemployed (this is on the assumption that the last five hours have the same productivity as the preceeding forty-four) than there actually were. Even though this means of calculation may yield a gross overestimate, it is still clear that this is an important way in which gorwth may reduce unemployment.

So even though job-sharing in a zero-growth economy may not be feasible, since it requires a negative marginal utility of income (or to put it a different way, an implausible degree of altruism in the work-force), in a growing economy, the diminishing marginal utility of income may nevertheless lead to something that is in effect job-sharing.

The third of the mechanisms may amplify the second. If economic growth is produced, as we have suggested in the earlier sections of this chapter, by a wave of social innovations, which involve the growth of new areas of informal work, then the marginal utility of money income may decline further. The argument is again derived from the model in Chapter 3. Informal economic activities become more productive, so work is transferred from the formal to the informal sector. As in the previous case, time outside the formal workplace becomes more valuable (in this case for another sort of work rather than for leisure) so we may be willing to take an even larger part of the increased productivity of the formal economy as non-paid-work time rather than money income. (The consequences in this case however may not be entirely positive, since, as we observed in Chapter 9, informal work tends to be considered as largely women's work. So it could be that part of the mechanism here would be specifically women sharing part of the formal employment with men, but not vice versa, particularly given the nursing and child-care examples of informal work suggested in the previous section.)

The general message from this brief discussion is that *if* we can manage to engineer a new wave of social innovation, by the sorts of technical and organizational changes discussed here, there may be some grounds for optimism about employment; this, partly as a result of the direct job-generating effects of economic growth, partly as a result of the job-sharing effects of the diminishing marginal utility of income, and partly as a result of the similar consequences of the informalization of production. We might note that social innovation has a dual impact. It increases the demand for labour, in the formal economy by establishing new markets for the new products used in the 'innovative modes of production', and also in the informal sector. And it reduces the supply of labour to the money economy.

10.6 Three Scenarios for the Future of Work

We must recognize that trends of events are in part determined by a logic that is internal to our social institutions and material environment, and quite out of our control. But nevertheless, the view propounded in this book is that the course of development is to some extent malleable. The future is affected by what we do. The following paragraphs will outline three stylized alternative future courses of events.

Scenario 1: Do nothing

If we do nothing, but simply continue (as in the UK) to try to hold down public expenditure in spite of the increasing cost of welfare payments to the unemployed, then it seems to be widely agreed among economic forecasters that unemployment will continue to rise. Particularly hard hit will be women, the young, inhabitants of regions and localities which have lost uncompetitive or otherwise outmoded industries, the less skilled, and members of racial minorities. There may be a growth of 'marginal' employment in low-skilled, very badly paid service occupations – a trend which might ameliorate the economic problems of the most disadvantaged groups, but with costs to their social and self-esteem, and without providing them with any long-term security. Associated with this trend we might expect a growth of the 'hidden' economy, particularly in the provision of personal and domestic services; again, this source of income is extremely insecure, and dependent on the existence of local pockets of prosperity to provide a market for its output. And an additional disadvantage of the hidden sector is that since it is by definition outside the law, those who work in it, who may already feel themselves excluded from the mainstream of life of the conventionally employed majority, will feel themselves further marginalized.

The generally depressed state of the economy does not mean that there is no growth in labour productivity in manufacturing industry. On the contrary, if firms are to stay in existence they will inevitably find themselves under the necessity of productivity growth. So the 'productivity gap' argument remains valid; the steady-state cost of public welfare provisions will continue to rise. This, in combination with the rising cost of income transfer payments to the unemployed, will inevitably mean further cuts in the output of welfare provisions. The already disadvantaged become even more disadvantaged.

This is, in all probability, not a stable development path: in the face of these mounting problems, governments cannot simply continue to do nothing. What will they do? We can identify two possible sorts of reactions, whose consequences are summarized as the other two scenarios.

Scenario 2: Protectionism

In the second scenario the government takes essentially defensive action to protect employment. It controls the import of manufactured goods to stop the haemorrhage of manufacturing jobs to foreign producers, and it increases public expenditure on the welfare state, again with the intention of maintaining employment. There are two points of view about the consequences of protecting manufacturing employment in the UK. Optimistic proponents of the protectionist argument claim that the shelter from competition will enable (because of the increased scale of markets) very considerable improvements in the productivity of manufacturing industry, which will enable the eventual dismantling of the protection at the point when British industry can successfully compete against foreign producers.

Sceptics (including the present writer) on the other hand feel that this ignores the roots of British competitive failure in the shortcomings of its management and union structures. Rather than encouraging faster productivity growth, this second viewpoint would suggest that protectionism would simply have the effect of feather-bedding the existing bad practices. Productivity growth would stay low, manufacturing industry would become progressivly less competitive – and so rather than being reversible, the need for protectionism to preserve employment would increase over time.

Admittedly, this sort of policy might be successful in its primary object of stopping the loss of jobs. But we must consider its social and political consequences. A country such as Britain following this policy on its own would have slower growth rates than other developed countries, which might lead to a need for emigration controls on skilled workers. The relatively low growth rate, in combination with increased taxation to pay for the public service sector would probably lead to inflationary pressures. These would in turn lead to a need for incomes and hence for prices policies – not as a temporary expedient, waiting for better times round the corner; but as a permanent fixture, backed by legal powers. And as the inflexibilities (shortages, quotas) inherent in such a system take hold, so we would expect the development of another sort of hidden economy, in its essence more sinister than that of the first scenario, dealing in black-market and smuggled goods; this would in turn lead to stronger police powers, and increased coercion of individuals by the state: in short, an increasingly authoritarian, illiberal, and undemocratic society.

Of course, if all the countries of the developed world were to follow such policies simultaneously, these processes might happen more slowly, and the end circumstances might be less extreme. But in fact this option is only really attractive to the less successful of the developed economies. The most likely eventuality is that some of the poor performers among the developed world will follow the protectionist route, ending up perhaps in situations similar to those of the more successful economies of Eastern Europe, while the rest adopt the policies which lead to the third scenario.

Scenario 3: Reconstruction

The policies leading to the third scenario bear some similarities to those leading to the second. Again they involve increased public expenditure in order to generate jobs, and the possibly inflationary consequences of the increased public expenditure are again avoided by a prices and incomes policy. Indeed, at the heart of this scenario there might be an historic political compromise between the political left and right, a new and more general 'social contract' whereby the right agree to more public spending in exchange for the left's agreement to controls to prevent wage-led inflation.

The crucial difference between the scenarios lies in the nature of the public expenditure. In the second scenario, markets are protected, and the public expenditure is devoted directly to the generation of jobs, without reference to their productivity. By contrast, in the third scenario, the public expenditure is specifically aimed at improving the overall productivity of the economy.

In this scenario, the increased public expenditure is directed to two areas. The first is investment in the telecommunications and other infrastructure necessary to enable and encourage the 'information society' social innovations discussed in Section 10.3. The precise nature of this investment is really behond the scope and competence of this book; for the moment we need simply say that its intention is to provide the conditions necessary for the development of new markets, both for household telecommunications, computing and information storage equipment, and, most important, for the software that will enable these to produce desirable services.

But as Section 10.4 suggested, these developments on their own might have a deleterious effect on the welfare state. If we wish to defend and enlarge public provision of services, we will also have to take action to make this provision more efficient. If the public is sceptical of the ability of public services to improve their productivity, simply increasing public

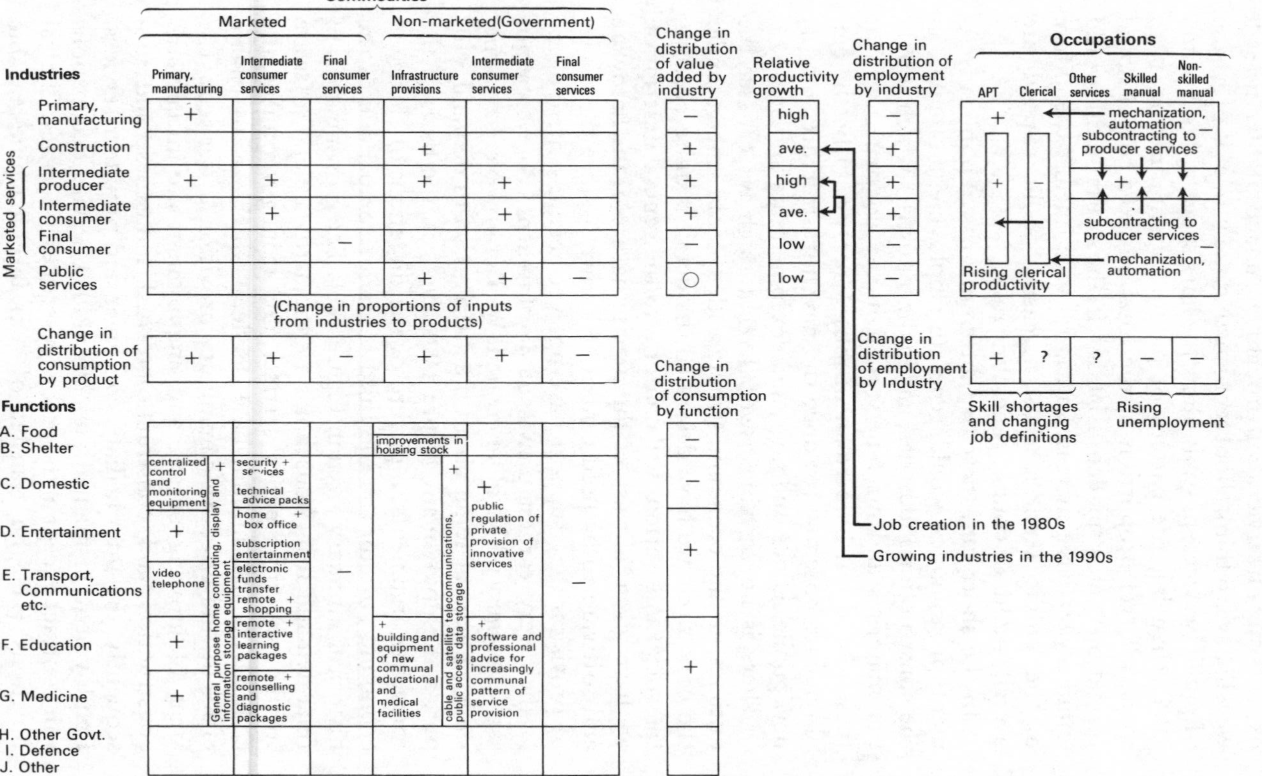

FIGURE 10.4 *Change in distribution of consumption, output, and employment, 1980s and 1990s*

expenditure on services will probably be unacceptable. Instead, we will have to think about reorganizing the mode of provision of public services, possibly along the lines suggested in Section 10.4; changing the pattern of public expenditure, so as to spend a larger proportion of the total on fixed costs (capital formation) and a smaller on labour. In the short run this would entail a very large increase in capital investment expenditure in the public services, thus increasing public expenditure overall – and, of course, generating jobs.

In the short run, these two different sorts of public investment programmes would generate employment directly, in the construction industries, and in the industries which supply the materials and machinery used in the construction. But these would not be permanent jobs. What would be the employment consequences once the new infrastructure, and the new public service buildings, and equipment are in place?

There is no certain answer. Even though we can see the outlines of the economic structure that could emerge under the conditions we have described, its detail is quite obscure; the social innovations will conform to the general description we have given, but as yet the precise nature of the hardware and software to be used cannot be known – since the modes of provision they support have not yet been invented, or are in their very earliest phases of development. So we cannot say *how many* jobs will be created. We can however say something in general terms about the *distribution* of demand, output, and employment. Figure 10.4 gives, on the basis of the foregoing arguments, a picture of the directions of change in economic structure in the third scenario, comparable to the historical summary given in Figure 10.1. The growing areas of final demand from households are the hardware and software for the new 'informatics' services, and from government, the equipment and intermediate services which go to enable efficient and effective community provision of services. The areas of growing output and employment are initially the construction and construction-related industries, and subsequently the intermediate consumer and producer services. Employment in manual and less skilled service occupations will contract except in the producer services industries. Clerical functions (i.e. the transfer of text into machine-readable form) will also be reduced. But will the growth in demand for administrative technical and professional workers be sufficient to offset the fall in other categories? This is the unanswerable question.

It is, however, not entirely implausible that the two processes for reducing the labour supply discussed in Section 10.5 (the effects of diminishing marginal utility of income and transfers of work to the informal economy), in combination with a growth in demand for labour from the intermediate service industries, might lead to some sort of balance. Though even in the most optimistic view there are still some serious problems. Any new demand for labour will be for the skilled, not the unskilled. And the growth of informal production is likely (particularly in the absence of positive steps to prevent this) to reduce women's paid work disproportionately. But nevertheless, in this scenario there is certainly some hope, both of economic growth and of new jobs, which is not present in the others.

10.7 The Revised Sequence

Emerging from the arguments of this book is a very substantial revision of the conventional view of the process of development of developed societies. Chapter 2 outlined the conventional view, of the 'march through the sectors', which gives us an apparently clear and unambiguous account of the sequence of development of the sectors of the economy. In Figure 10.5 we see the traditionally accepted version in which the primary,

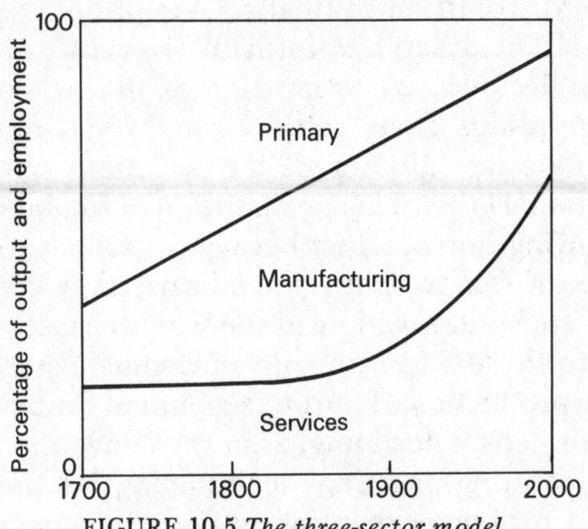

FIGURE 10.5 *The three-sector model*

secondary, and tertiary sectors succeed one another as the focus of economic activity. We do not know, however, whether these 'sectors' refer to consumption categories, industrial output categories, industrial employment categories, or occupational categories. And we have found, in the course of the arguments in the foregoing chapters, that the relationships between these various classificatory dimensions are by no means simple. The 'service' categories in the consumption dimension, for example, have no necessary relation to the 'service' categories in, say, the occupational employment dimension. So the conventional wisdom is by no means as clear and unambiguous as it may at first appear. But what can we replace it with?

The arguments of this book establish, in effect, an alternative to the 'march through the sectors', a revised sequence of development. It is more complicated than the conventional view – necessarily so, for it deals separately with the classificatory dimensions confounded in the simpler approach.

The six graphs in Figure 10.6 show the pattern of development, for some idealized developed economy, envisaged in this revised sequence. In the 'distribution by final service function' graph we see the 'Engel's Law' hierarchy of needs reflected in differential rates of growth of consumption of commodities associated with the various final functions. This pattern of succession from 'basic' to 'luxury' functions is not however reflected by a growth of final demand for services, because of change in the mode of provision for these services. We see that in the 'distribution of final consumption' graph, demand for final marketed services initially rises, but after 1950 or thereabouts declines, as a proportion of all final consumption. This change results from 'self-servicing' types of social innovation. By the last quarter of the twentieth century we see the proportional growth in consumption of final non-marketed services slowing down, either because of taxpayers objections to increases in welfare state expenditure (as in scenario 1), or because of social innovation in the welfare state (scenario 3).

Moving to the 'distribution of value added by sector' graph, we find the proportion of output accounted for by the primary sector continuously declining, as in the conventional account, but manufacturing industry maintaining its proportion of total output for longer than conventionally expected, because

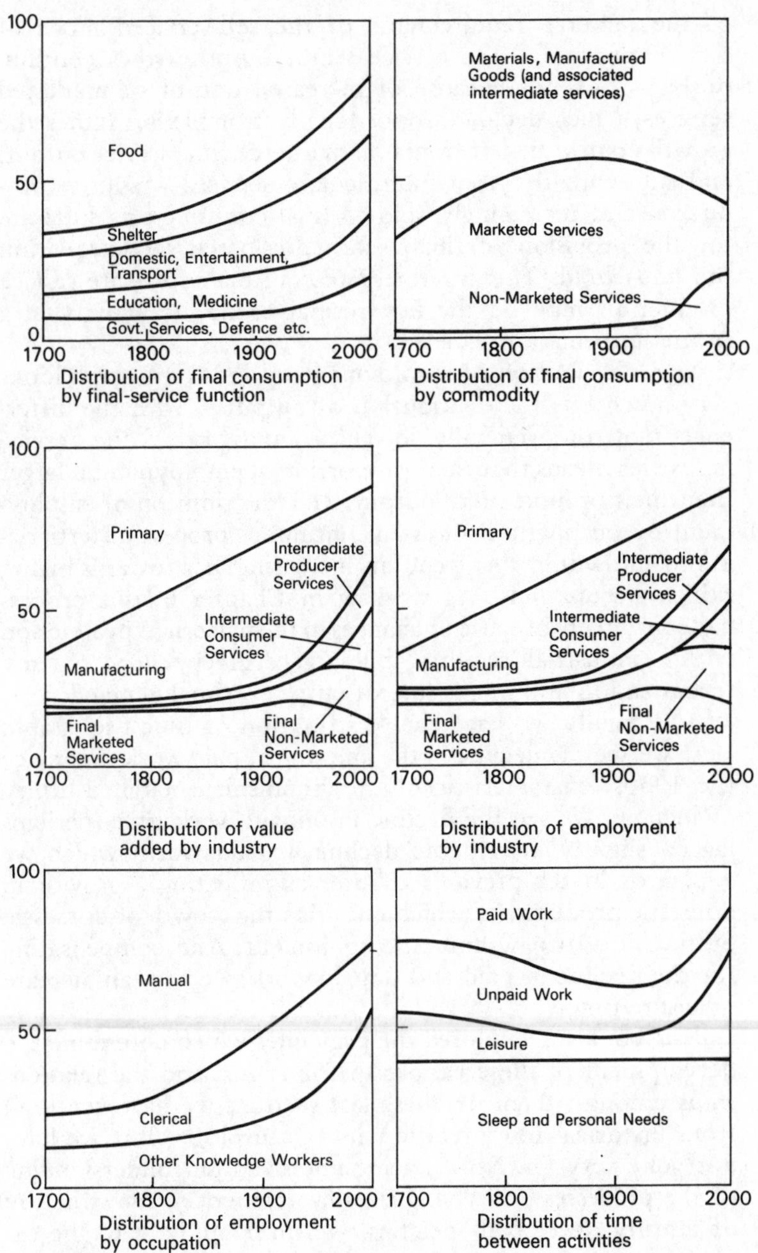

FIGURE 10.6 *Schematic development of a typical European developed society*

of the material requirements of the self-serviced mode of service provision. The service sector output grows continuously – but not because of increased output of marketed services, which decline proportionally from 1950; rather the growth comes, at first from non-marketed final service output, and subsequently from intermediate services – sold to producers, and increasingly sold to final consumers as software for the provision of final informatics-based services within the household. This intermediate consumer software service production may be the key formal economic activity in a future 'information society'.

The 'distribution of employment by industry' graph looks very much like the distribution of output – with the difference that the generally lower productivity of the service industries means that their proportion of employment is larger than their proportion of output. The 'distribution of employment by occupation' shows a continuous process of tertiarization, and within this trend an increasing bias towards higher skilled occupations. As we saw in Chapter 6, this process happens largely because of changes in the pattern of production within industrial sectors, and largely independent on any tertiarization in the structure of output or final demand.

And finally we have the 'distribution of time use' graph. Here we see the decline in the amount of paid work time since the 1950s, characteristic of the diminishing marginal utility of income. We see the decline in unpaid work time (perhaps lagged slightly behind the decline in paid work) which we explained in the previous chapter as reflecting a growth in domestic productivity which outstrips the growth of domestic output due to new domestic equipment. And compensating for the decline in paid and unpaid work, we have an increase in leisure time.

This book has explored the principles which determine the development of these various distributions, and the relationships amongst them. In these last sections we have ventured from historical interpretation into futurology. But we have not gone very far. Now, having improved our understanding of the processes underlying the development of the structure of already developed societies, we must apply it to the exploration of the possible futures that lie ahead of us: this is the aim of the next phase of our research project.

Afterword

THE foregoing represents the current state of our work. As always, research begets the need for more research. Three topics emerge as being of critical concern:

1. *The design of the future telecommunications infrastructure*. It is quite clear that the sorts of social innovation that are possible – and hence the nature of the life-style and employment implications of the new technologies – are critically dependent on some design and related issues. What is the desirable general bandwidth (or bandwidths) of the telecommunications system? The answer will determine in particular whether households can use the new infrastructure to receive television and other high bandwidth video signals. Will all terminals to the communications system have, in principle, equal access to all other terminals (a 'symmetrical' system), or will the system be designed to enable some terminals to transmit and receive more complex information than others (a 'hierarchical' system, along the lines of interactive cable television)? This will determine the ease with which very small 'information provider' firms can be set up – and also determine the pattern of formal economy work done in the home. What will be the legal situation for copyright on software, and for privacy of personal information? These issues will determine among other things the level of investment in software. Clearly there will be some need for public control to ensure that the right design decisions will be adopted – but does this in fact call for public investment, or for publicly regulated private investment along the lines of the current USA telephone system?

2. *The redevelopment of the welfare state*. It follows from our argument that the maintenance, let alone the improvement, of services provided by the welfare state, may require some very substantial organizational and technical innovation. How can we promote these innovations – and can we do so *and* protect jobs, particularly skilled jobs? How do we organize the informal communal groups which may increasingly take responsibility for final service provision, and what will be their relationship with the formal economy institutions in the same field? And underlying these questions is a broader issue: how do we organize what is almost inevitably a bureaucratic sector of the economy so that it does not repress efficiency-promoting innovations?

3. *The distribution of unpaid work.* Even if our optimistic Scenario 3 does work out successfully, it still only gives benefits in the form of the number of paid jobs; it does nothing about the distribution of jobs between men and women. If, within this scenario, the amount of informal production is growing, and the sexual segregation of unpaid work remains unchanged, the consequence will inevitably be further pressure on women's time, further prejudicing their ability to hold down rewarding paid employment. What sorts of changes in employment legislation, tax distributions, family legislation, children's and old people's care provisions, would help to remedy this situation?

Index